W9-CSA-903

Microwave-Assisted Proteomics

Dedication

To my wonderful husband Glen and my beautiful son Joseph.

Microwave-Assisted Proteomics

Jennie Rebecca Lill
Genentech Inc., South San Francisco, CA, USA

RSCPublishing

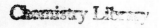
QP
551
L55
2009
CHEM

ISBN: 978-0-85404-194-7

A catalogue record for this book is available from the British Library

© Jennie Rebecca Lill, 2009

All rights reserved

Apart from fair dealing for the purposes of research for non-commercial purposes or for private study, criticism or review, as permitted under the Copyright, Designs and Patents Act 1988 and the Copyright and Related Rights Regulations 2003, this publication may not be reproduced, stored or transmitted, in any form or by any means, without the prior permission in writing of The Royal Society of Chemistry or the copyright owner, or in the case of reproduction in accordance with the terms of licences issued by the Copyright Licensing Agency in the UK, or in accordance with the terms of the licences issued by the appropriate Reproduction Rights Organization outside the UK. Enquiries concerning reproduction outside the terms stated here should be sent to The Royal Society of Chemistry at the address printed on this page.

Published by The Royal Society of Chemistry,
Thomas Graham House, Science Park, Milton Road,
Cambridge CB4 0WF, UK

Registered Charity Number 207890

For further information see our web site at www.rsc.org

Preface

Since the conceptualization of the electromagnetic spectrum and development of the magnetron, microwave energy has been utilized in many aspects and disciplines of science. Although adopted by multiple industries over the past quarter of a century, it is only within the past few years that microwave irradiation has been evaluated as a useful tool for the biochemical and chemical preparation of proteins and other biomolecules for proteomics and in particular mass spectrometric analysis. This book describes the evolution and integration of microwave energy into the biosciences with particular emphasis on the proteomic arena. An in-depth evaluation of a variety of techniques within the field of proteomics that benefit from microwave irradiation is given. This book chronicles the development of these microwave-assisted methods and provides a synopsis of the final protocols that have become standardized for each area discussed. This book also focuses on the types of instrumentation that may be employed for microwave-assisted protein chemistries and the hypotheses of mechanisms of action for the microwave-enhanced methodologies.

Although still in its infancy, the application of microwave assistance is gaining momentum in several fields, in particular that of proteomics and protein chemistry. A single monograph cannot comprehensively cover this rapidly evolving field; however, this publication should provide an in-depth introduction to the science of microwave-assisted proteomics and will hopefully ignite some ambition within researchers interested in trying these protocols in their own laboratories.

Jennie R. Lill PhD

Microwave-Assisted Proteomics
By Jennie Rebecca Lill
© Jennie Rebecca Lill 2009
Published by the Royal Society of Chemistry, www.rsc.org

Contents

Microwave-Assisted Proteomics
By Jennie Rebecca Lill
© Jennie Rebecca Lill 2009
Published by the Royal Society of Chemistry, www.rsc.org

Acknowledgements

I would like to thank all of my team, in particular Wendy Sandoval, Victoria Pham, Peter Liu, Elizabeth Ingle, Dr David Arnott and Dr Richard Vandlen for insightful discussion and advice during this project. Thanks go to Allison Bruce for designing the graphics and to Daniel Burdick and Mark Laiktham for introducing us to the world of microwave assistance. A big thank you is due to all the researchers who pioneered this field and initiated the excitement in our group regarding microwave-assisted proteomic technologies. I thank my parents, family and friends for their support and Poppy and Splodgey for understanding that my weekends have become more about book writing than beach activity! A final thank you to Dr Merlin Fox and the Royal Society of Chemistry for inviting me to write this monograph; it has been a truly educational and exciting endeavor.

CHAPTER 1

Evolution of Microwave Irradiation and Its Introduction to the Biosciences

Abstract

Since the conceptualization of the electromagnetic spectrum, through the development of the magnetron microwave energy has been utilized in many aspects and disciplines of science. Although adopted by many industries over the past quarter of a century, it is only within the past few years that microwave irradiation has been evaluated as a useful tool in the biochemical and chemical preparation of proteins and other biomolecules. This chapter describes the evolution of the magnetron and some early applications of microwave assistance in the bioanalytical sciences.

1.1 Microwave Radiation

The electromagnetic spectrum is a continuum of all electromagnetic waves arranged according to frequency and wavelength. Microwaves occupy the electromagnetic spectrum between infrared radiation and radio waves and have wavelengths between 0.01 and 1 m with a frequency range between 0.3 and 30 GHz (Figure 1.1). Most commercially available microwaves have a narrower range at around 2.5 GHz.[1] Microwave energy is a natural phenomenon which can be induced when electric current flows through a conductor, for example an antenna, a transmitter chip or a magnetron.

Microwave-Assisted Proteomics
By Jennie Rebecca Lill
© Jennie Rebecca Lill 2009
Published by the Royal Society of Chemistry, www.rsc.org

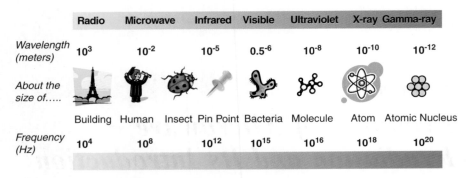

Radio	Microwave	Infrared	Visible	Ultraviolet	X-ray	Gamma-ray

Wavelength (meters): 10^3 10^{-2} 10^{-5} 0.5^{-6} 10^{-8} 10^{-10} 10^{-12}

About the size of.....: Building Human Insect Pin Point Bacteria Molecule Atom Atomic Nucleus

Frequency (Hz): 10^4 10^8 10^{12} 10^{15} 10^{16} 10^{18} 10^{20}

Figure 1.1 The electromagnetic spectrum modified from http://en.wikipedia.org/wiki/Electromagnetic_spectrum. Microwaves reside in between radio waves and the infrared portion of the electromagnetic spectrum.[2]

1.2   History and Evolution of the Magnetron

The theory of microwave irradiation was first documented in 1864 by James Clerk Maxwell, a Scottish mathematician and theoretical physicist. In the eponymous Maxwell's equations he described a unified model for electromagnetism and paved the way for modern physics.[3] It was not until over twenty years later that microwave irradiation was physically demonstrated by Heinrich Hertz.[4] Hertz was a German physicist who demonstrated the existence of electromagnetic radiation by building antennae and apparatus to produce and detect high-frequency radio waves. Microwaves were produced using various apparatus up until the 1920s when Albert Hull, a researcher at General Electric's research laboratories, invented the simple two-pole magnetron, or split-anode magnetron.[5] Although revolutionary at its time, the two-pole magnetron was relatively inefficient and was soon superseded by the resonant-cavity magnetron which proved to be more efficient and convenient.

The first half of the twentieth century become synonymous with large-scale war and scientific innovation, the combination of which led to the development of many technologies including the introduction of radar. In a deal between British and American researchers, the cavity magnetron was developed into a viable radar system, and by 1941 magnetrons for radar systems were being manufactured at a rate of 17 per day at Raytheon. It was during this time that a researcher at Raytheon, Percy Lebaron Spencer, made two important discoveries. Firstly, he was awarded the Distinguished Public Service Award by the US Navy for significantly improving the manufacturing process of magnetrons and increasing production more than 100-fold. Secondly, perhaps more famously, in 1945 while standing in front of an open magnetron he noticed that a chocolate bar had melted in his pocket. After several other "tests" including popping popcorn and exploding eggs he concluded that microwave radiation could be tailored for use in cooking devices and hence the invention of the microwave oven.[6,7]

By 1947 the first commercial microwave oven had been manufactured by Raytheon, although during the first few years of commercialization these ovens stood at nearly 6 feet tall and weighed over 700 pounds. By the 1970s microwave ovens had become much more accommodating for household use, and by 1975 the sales of microwave ovens started to exceed those of gas oven ranges in the USA.

In 1978 the first commercial microwave for laboratory use was introduced by CEM and, since then, laboratory microwaves have increased in sophistication and utility to include models specific for the chemical and biological sciences. Figure 1.2 shows the time scale of the evolution of the magnetron.

1.3 Microwaves as a Catalysis Tool in Organic and Inorganic Chemistry

In 1986 the first reports of high-speed chemical synthesis with microwave assistance were published.[8] Since then there have been numerous publications describing microwave-assisted synthetic reactions where most researchers observe shorter reaction times, increased yields and cleaner syntheses due to reduced by-product formation or side reactions (for a comprehensive review, see Alcazar *et al.*[9]). Originally, chemists would employ microwave assistance only for those reactions that proved troublesome or that resulted in poor yields. Nowadays, however, as instrumentation and an understanding of the mechanisms involved have matured, chemists routinely employ microwave-catalyzed protocols at the first stage of method development. Indeed for reactions involving highly polar reagents or metal catalysis, microwave irradiation is confirmed as the most valuable mode of heating available, and standard protocols for polymer synthesis and process control typically employ microwave assistance.[10–12]

1.4 Microwave-Assisted Staining of SDS-PAGE and PVDF Membrane-Embedded Proteins

One of the first microwave-assisted applications in the biological setting was the fixing, staining and destaining of sodium dodecyl sulfate polyacrylamide gels and poly(vinylidine difluoride) (PVDF) membranes. Sodium dodecyl sulfate polyacrylamide gel electrophoresis (SDS-PAGE) is a highly valuable technique employed in biochemistry, genetics and molecular biology to separate biomolecules according to their electrophoretic mobility. In the separation of proteins, the electrophoretic mobility is dictated by the length of the polypeptide chain or molecular weight as well as higher order protein folding, posttranslational modifications and other factors. After proteins are separated on the gel, the sample is typically fixed with a reagent to mobilize the gel and to stop migration or dispersion. Fixation is typically performed in a high percentage of methanol,

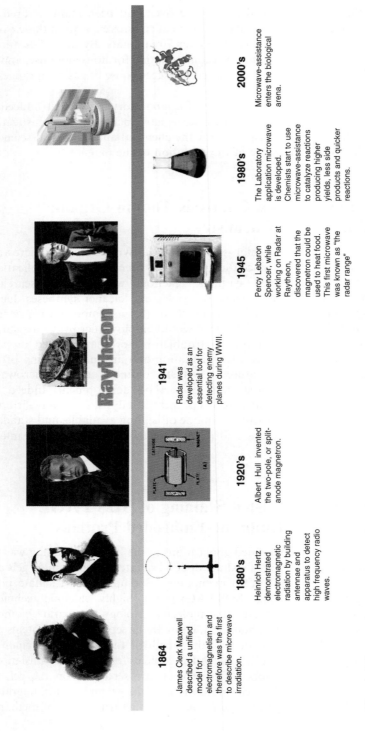

Figure 1.2 A time line of important events in the history of microwave energy discovery through its evolution as a tool for analytical biosciences.

1864

James Clerk Maxwell described a unified model for electromagnetism and therefore was the first to describe microwave irradiation.

1880's

Heinrich Hertz demonstrated electromagnetic radiation by building antennae and apparatus to detect high frequency radio waves.

1920's

Albert Hull invented the two-pole, or split-anode magnetron.

1941

Radar was developed as an essential tool for detecting enemy planes during WWII.

1945

Percy Lebaron Spencer, while working on Radar at Raytheon, discovered that the magnetron could be used to heat food. This first microwave was known as "the radar range"

1980's

The Laboratory application microwave is developed. Chemists start to use microwave-assistance to catalyze reactions producing higher yields, less side products and quicker reactions.

2000's

Microwave-assistance enters the biological arena.

which can clean up the gel from any remaining material from the SDS running buffer. After fixation the gel is stained using one of the many stains available, for example Coomassie Blue or silver stain. Traditional staining protocols recommend immersion of the gel or membrane into stain solution for many hours (often an over night incubation). After staining, gels are destained to remove background stain, and to allow the bands corresponding to the proteins of interest to be visualized. Proteins may also be electro blotted from the gel onto PVDF membranes whereby the sample is more compatible with long-term storage and with certain analytical techniques, for example Edman degradation. PVDF membranes can be stained and destained in the same manner as SDS-PAGE.

Microwave irradiation has been successfully employed to speed up the process of fixation, staining and destaining since the early 1990s; however, the first citation on record appears to be that of Nestayy *et al.* Here an in-depth study into the effect of microwave irradiation on the staining of proteins in gels or membranes using a variety of common stains was performed.[13] Microwave-assisted staining with Coomassie Blue, SYPRO® Ruby, silver stain and colloidal gold protocols was evaluated. Nestayy *et al.* demonstrated that the traditionally time-consuming process of staining and destaining gels could be significantly reduced if performed in the presence of microwaves. A regular domestic microwave oven was used and gels were introduced into the cavity of the microwave in Petri dishes or any microwavable container. It was postulated that the faster staining and destaining was mainly due to heat produced by the microwaves which maximized the efflux and influx of solvent and solutes from the gel or membrane.

After separation of proteins by SDS-PAGE, proteins were identified by performing an in-gel tryptic digestion followed by tandem mass spectrometric characterization (see Chapter 4). Nestayy *et al.* went on to monitor the effect microwave-assisted staining had on subsequent mass spectrometric analysis. Increased proteolytic cleavage was observed after microwave-assisted staining compared to conventional methods (*i.e.* room temperature incubation). It was proposed that this effect was due to increased denaturation of proteins embedded in the matrices of the gel or membrane after microwave exposure. (Note that the digestion itself was not performed in the presence of microwave radiation, only the staining and destaining of the gel or membrane.) This denaturation resulted in greater accessibility of the substrate proteolytic sites to the enzyme. Heat generated from the microwave process may also have contributed to gel or membrane expansion, therefore enhancing exposure of the protein to enzymatic cleavage, and also resulting in increased extraction of peptides from the gel or membrane after digestion.

Overall it was concluded that microwave irradiation of proteins separated by SDS-PAGE or blotted onto PVDF membranes often significantly improved proteolytic coverage when compared to traditional gel-staining techniques. In addition, there did not appear to be any detrimental effects such as loss of posttranslational modifications or increased deamidation or oxidation. For a practical protocol on microwave-assisted fixation, staining with Coomassie Blue and destaining of gels, refer to protocol I in Chapter 10.

A microwave-enhanced ink staining method was also recently reported whereby dye-based blue-black ink was used to quantitatively visualize proteins spotted onto a nitrocellulose membrane by incubation in a domestic microwave oven. The total staining time was reduced from more than 30 min to less than 3 min by employing microwave assistance. A 500-fold dynamic range from low nanogram to mid-microgram total protein amounts could be detected using this method, which in addition allowed samples in complex buffers and chaotropes to be quantified.[14]

1.5   Microwave-Assisted Peptide Synthesis

Peptides may be synthesized for a number of reasons including internal standards in mass spectrometric quantitation studies, synthetic hormones or neurotransmitters, for mapping enzyme specificity or drug interaction sites and other bio-pharmacological tools. Chemical synthesis of peptides is often preferred to *in vivo* or *in vitro* synthesis as samples are free from other cellular debris such as contaminating peptides, lipids or genetic material. In addition, chemical synthesis allows flexibility in the design of the peptide and allows incorporation of specific modifications or isotopic and other non-natural amino acids into the peptide backbone.

Microwave irradiation as a tool for peptide synthesis was first described in 1992,[13] and since then has been applied to thousands of peptides, some composed of up to as many as 200 amino acid residues. Microwave-assisted peptide synthesis has many advantages over conventional solution- and solid-phase protocols including higher yields, higher specificity during coupling, higher deprotection and less racemization.[15] Although the exact mechanisms of microwave assistance are not proven, the main hypothesis is driven by evidence that, during synthesis using conventional methods, the reaction matrix can be rendered inaccessible leading to aggregation with both itself and neighboring peptides. Microwave irradiation may lead to deaggregation *via* dipole alignment, allowing increased accessibility of reagents and hence more efficient deprotection, coupling and washing.[16,17]

1.6   Microwave-Assisted Antigen Retrieval

Immunohistochemistry or immunohistochemical staining is the process whereby tissues, cells or proteins are localized using antibodies raised against antigens of interest. Typically this is accomplished using either a primary antibody or combination of a primary and secondary antibody that is conjugated to a visualization tool, for example a fluorophore or fluorescent molecule (refer to Chapter 8 for more microwave-assisted immunoassays). Many samples, for example tumor biopsies, requiring immunohistochemical staining are fixed in formalin or paraffin embedded for preservation, treatments which are not conducive to antigen exposure. To increase antigen exposure (a protocol known as antigen retrieval), microwave irradiation is commonly applied. Microwave irradiation using a domestic microwave oven can increase antigen exposure, allowing antibodies to have increased binding capabilities to target antigens. Many in-depth studies into microwave-assisted antigen retrieval and fixation of tissues have been reported.[18-21] Immunohistochemical protocol incubation times can be significantly reduced if the sample is pulsed with short bursts of microwave exposure, either prior to or during incubation of the sample with the primary antibody. For a practical protocol on microwave-assisted antigen retrieval, refer to protocol II in Chapter 10.

1.7   Microwave-Assisted Analysis of Neuropeptides

Neuropeptides are involved in a variety of physiological and pathophysiological functions including metabolism, memory and sensory perception, regulation of appetite, learning and mood.[22,23] Neuropeptides are often present at significantly lower levels compared to degradation products of other more abundant proteins from neuronal tissue. Therefore, the field of neuropeptidomics is extremely challenging due to this wide dynamic range. Analogous to the applications mentioned in Section 1.6, microwave irradiation has assisted in the stabilization and fixation of neuronal tissue for immunohistochemistry.[24] Microwave assistance allowed analyses to be approximately 10^3 to 10^4 orders of magnitude faster than the conventional immersion fixation method.[25] Incorporation of microwave-mediated fixation in conjunction with microscale sample deposition onto matrix-assisted laser desorption ionization (MALDI) target plates has allowed the detection of neuropeptides at amounts as low as 80 attomoles using MALDI time-of-flight mass spectrometry.[26]

1.8   Microwave-Assisted Akabori Reaction

In the 1950s two techniques were developed that allowed analysis of either the N-terminus (Edman degradation[27]) or C-terminus of a protein or peptide. The Akabori reaction allows identification of the C-terminus of peptides or proteins by heating in the presence of anhydrous hydrazine.[28,29] This method remained a cumbersome technique until Bose *et al.* modified the reaction by employing microwave energy using a domestic microwave oven[30] by demonstrating that this conventionally overnight reaction could be performed in minutes. In addition to reducing incubation times, microwave irradiation could be performed in an open vessel whereas the conventional technique required samples to be heated in a sealed tube. The report of the microwave-assisted Akabori reaction was one of the preliminary papers to demonstrate the fundamentals of microwave chemistry applied to a bioanalytical protocol.[31] The microwave-mediated Akabori reaction also proved useful in the analysis of cyclic peptides, as here cyclic peptides could be cleaved in a selective manner to produce mainly the hydrazide product which could then be analyzed by mass spectrometry to gain full peptide sequencing information.[32]

References

1. D. M. Pozar, *Microwave Engineering*, John Wiley, New York, 2nd edn, 1997.
2. http://en.wikipedia.org/wiki/Electromagnetic_spectrum
3. J. C. Maxwell, *An Elementary Treatise on Electricity*, Clarendon Press, 2nd edn, Oxford, 1888.
4. J. Z. Buchwald, *The Creation of Scientific Effects: Heinrich Hertz and Electric Waves*, University of Chicago Press, Chicago, 1994.
5. *Electronic Design for Engineers and Engineering Managers*, Hayden Publishing, Rochelle Park, NJ, 1976, vol. 24.
6. T. J. Morgan, *RADAR*, F. Muller, London, 1960.
7. R. Buderi, *The Invention that Changed the World: How a Small Group of Radar Pioneers Won the Second World War and Launched a Technological Revolution*, Simon and Schuster, New York, 1996.
8. R. Gedye, F. Smith, K. Westaway, H. Ali and L. Baldisera, *Tetrahedron Lett.*, 1986, **26**, 279.
9. J. Alcazar, G. Diels and B. Schoentjies, *Mini Rev. Med. Chem.*, 2007, **7**, 345.
10. J. J. Chen and S. V. Deshpande, *Tetrahedron Lett.*, 2003, **44**, 8873.
11. C. E. Humphrey, M. A. M. Easson, J. P. Tierney and N. J. Turner, *Org. Lett.*, 2003, **5**, 849.
12. N. E. Leadbeater, S. J. Pillsburg, E. Shannahan and V. A. Williams, *Tetrahedron*, 2005, **61**, 3565.
13. V. J. Nestayy, A. Dacanay, J. F. Kelly and N. W. Ross, *Rapid Commun. Mass Spectrom.*, 2002, **16**, 272.
14. X-P. Wu, Y-S. Cheng and J-Y. Liu, *J. Proteome Res.*, 2007, **6**, 387.
15. H-M. Yu, S-T. Chen and K-T. Wang, *J. Org. Chem.*, 1992, **57**, 4781.
16. J-P. Tam and Y. A. Lu, *J. Am. Chem. Soc.*, 1995, **117**, 12058.
17. J. M. Collins and N. E. Leadbeater, *Org. Biomol. Chem.*, 2007, **5**, 1141.
18. M. K. Patterson Jr and R. Bulard, *Stain Technol.*, 1980, **55**, 71.
19. Z. A. Hafajee and A. S. Leong, *Pathology*, 2004, **36**, 325.
20. S. G. Temel, F. Z. Minbay, Z. Kahveci and L. Jennes, *J. Neurosci. Meth.*, 2006, **156**, 154.
21. L. L. Emerson, S. R. Tripp, B. C. Baird, L. J. Layfield and L. R. Rohr, *Am. J. Clin. Pathol.*, 2006, **125**, 176.
22. D. Krieger, *Science*, 1983, **222**, 975.
23. M. Konnig, A. Zimmer, H. Steiner, P. Holmes, J. Crawley, M. Brownstein and A. Zimmer, *Nature*, 1996, **383**, 535.
24. G. R. Login, S. J. Schnitt and A. M. Dvorak, *Eur. J. Morphol.*, 1991, **29**, 206.
25. G. R. Login and A. Dvorak, *J. Neurosci. Meth.*, 1994, **55**, 173.

26. H. Wei, S. L. Dean, M. C. Parkin, K. Nolkrantz, J. P. O'Callaghan and R. T. Kennedy, *J. Mass Spectrom.*, 2005, **40**, 1338.
27. P. Edman, *Eur. J. Biochem.*, 1967, **1**, 80.
28. S. Akabori, *J. Chem. Soc. Japan*, 1931, **52**, 606.
29. S. Akabori, K. Ohno and K. Narita, *Bull. Chem. Soc. Japan*, 1952, **25**, 214.
30. A. K. Bose, Y-H. Ing, H. Lavlinskaia, C. Sareen, B. N. Pramanik, P. L. Bartner, Y-H. Liu and L. Heimark, *J. Am. Soc. Mass Spectrom.*, 2002, **13**, 839.
31. J. R. Lill, E. S. Ingle, P. S. Liu, V. Pham and W. N. Sandoval, *Mass Spectrom. Rev.*, 2007, **26**, 657.
32. B. N. Pramanik, Y. Hain, A. K. Bose, L-K. Zhang, Y-H. Liu, S. N. Ganguly and P. Bartner, *J. Am. Soc. Mass Spectrom.*, 2003, **44**, 2565.

CHAPTER 2
Microwave Instrumentation in the Biosciences

Abstract

Many formats of microwave apparatus have been described in the literature for carrying out microwave-assisted protein chemistries and proteomic sample preparation. This chapter discusses some of the terminology commonly employed for describing the different types of microwave technology and discusses the options available to researchers in terms of deciding which type of system to purchase or develop for the needs of individual laboratories.

2.1 Background

Chapter 1 discussed the background of microwave irradiation and its evolution into a tool for the catalysis of chemical and biochemical reactions. In brief, the first true magnetron was invented by the researcher Albert Hull in the 1920s, who invented the simple two-pole magnetron, or split-anode magnetron.[1] This two-pole design was relatively inefficient and was soon superseded by the resonant-cavity magnetron which proved to be more efficient and convenient. Cavity magnetrons, the norm in microwave design today, were developed by Raytheon during the 1940s as a technology specifically for use in radar systems during World War II. It was during this time that Percy Lebaron Spencer, a researcher at Raytheon, was awarded the Distinguished Public Service Award by the US Navy for significantly improving the manufacturing process of magnetrons and increasing production more than 100-fold. However, more famous than the prestigious award with which he was bestowed was his discovery of the utility of microwave power as a tool for heating food. In 1945 while standing in front of an open magnetron he noticed that a chocolate bar

Microwave-Assisted Proteomics
By Jennie Rebecca Lill
© Jennie Rebecca Lill 2009
Published by the Royal Society of Chemistry, www.rsc.org

had melted in his pocket. Through this observation he concluded that microwave irradiation could be tailored for use in cooking devices and hence the invention of the domestic microwave oven.[2,3]

During the next 30 years the magnetron was further developed both as a military tool and for use in domestic microwave ovens, and in 1978 the first commercial microwave for laboratory use was introduced by CEM. Since then laboratory microwave ovens have increased in sophistication and utility to include models specific for chemical and biological science applications. During the 1980s, microwave-assisted chemistries were first demonstrated and many reactions were shown to work a thousand times faster when microwave irradiation was employed compared to utilizing conventional heating mechanisms. In the 1990s Milestone introduced the first high-pressure digestion vessel (model HPV 80) which helped complete the digestion of difficult compounds such as oxides, oils and pharmaceutical compounds. Also in the 1990s CEM developed a microwave digestion system (MDS 2000) with a batch reactor which increased throughput by allowing several analytical tests to be performed simultaneously. Many different commercial models are now available, tailored for utility with specific reactions or processes in mind. In addition, several academic groups have fashioned their own microwave systems to suit their specific research needs, several of which are discussed herein. In 2003 the market for laboratory-based microwave systems was estimated to be around $89 million, with CEM, Biotage and Milestone being the major players (although several other companies also produce excellent and more tailored instrumentation). This chapter summarizes the key elements of the different types of microwave apparatus and demonstrates the utility of several marketed and academic instruments.

2.2 Microwave Design

2.2.1 Monomode *vs.* Multimode Systems

Monomode (sometimes referred to as single-mode) microwave apparatus are characterized by the ability to create a standing wave pattern which is generated due to interfering fields of the same amplitude but with differing oscillating directions. As demonstrated in Figure 2.1, an array of nodes (where the microwave energy intensity is zero) and antinodes (where the magnitude of the microwave energy is at its highest) is produced in a monomode microwave system.[4]

One important consideration when utilizing a monomode microwave system is the distance of the sample from the magnetron. The sample needs to be within an appropriate distance from the antinode of the electromagnetic wave path. During single-mode operation typically only one vessel (or cluster of small vessels) can be exposed to microwave irradiation at a given time (see Figure 2.2). Due to this limitation, monomode apparatus is typically employed for small-scale drug discovery and combinatorial chemistry applications.[4]

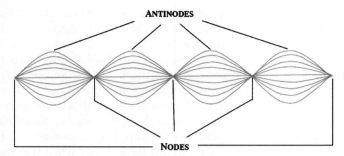

Figure 2.1   Generation of a standing wave pattern.[4] (Reproduced with permission from the American Chemical Society.)

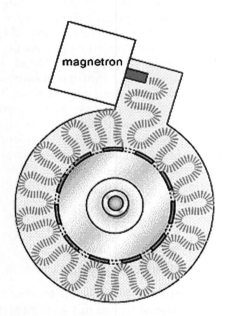

Figure 2.2   Schematic of a single-mode microwave heating source. (Reproduced with permission from CEM Corporation.)

Monomode microwaves, however, are advantageous where rapid heating is needed, which is due to the fact that the sample is always positioned at the apex of the antinodes, *i.e.* where the microwave field is at its most dense.

Multimode microwave systems differ from the monomode apparatus in that they do not generate a standing wave pattern and instead chaotic microwave dispersion is deliberately induced (see Figure 2.3). By inducing as much chaos as possible, an increased area can undergo effective heating, and as a result multimode microwave systems can accommodate a much larger number of

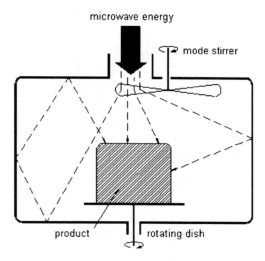

Figure 2.3 Schematic of a multimode microwave heating source.[4] (Reproduced with permission from the American Chemical Society.)

samples than monomode systems. Due to these salient features, the multimode microwave apparatus is typically employed for bulk/large-sale heating reactions. Domestic microwave ovens also typically operate using this mode of operation. One disadvantage of multimode microwave-assisted heating, however, is that temperature dispersion cannot be efficiently controlled and samples may be more susceptible to hot spots and uneven temperature distribution.

2.2.2 Continuous Flow Systems

In 2005 Comer and Organ described a system comprising of a continuous flow microwave-assisted parallel capillary which, although not designed with microwave-assisted proteomics in mind, did show the potential for flow-based systems which could potentially be utilized for proteomic experiments.[5] The design consisted of a capillary with an internal diameter ranging from 200 to 1200 μm, while the flow rates varied between 2 and 40 μL min^{-1} (which corresponded to the sample being irradiated for approximately 4 min). Figure 2.4 shows a schematic of the basic continuous flow reactor design. This design consisted of a stainless steel holding/mixing chamber with three inlet ports which merged into one outlet. The inlets were connected *via* micro-tight fittings and Teflon tubing to an external syringe pump. Capillary tubes of varying internal diameters (200–1150 μm) were interchangeably attached to the holder by micro-tight fittings. After exiting the reaction capillary, the reaction flowed *via* Teflon tubing directly to a monitoring device or collection vessel. The holder was positioned on top of the microwave cavity of a Biotage Smith Creator Synthesizer, allowing the capillary to be kept in place within the irradiation chamber. The capillary was irradiated with 2.45 GHz of single-mode

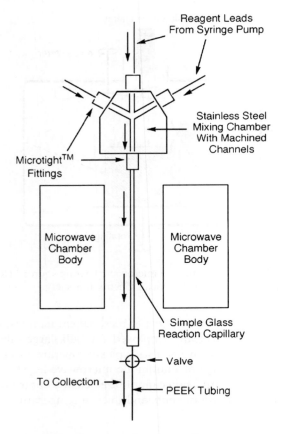

Figure 2.4   Schematic of a basic continuous flow reactor design (PEEK, poly(ether ether ketone)).[5] (Reproduced with permission from the American Chemical Society.)

microwave power that could be varied between 0 and 300 W, while the reaction temperature was monitored by an internal infrared sensor. Even at these low volumes, the reaction vessels were proven to be capable of picking up the effect of microwave irradiation. Comer and Organ showed the system was reproducible and despite using narrow capillaries, clogging of material, even that containing particulates, was not typically observed.

In 2008 Hauser and Basile described an online microwave system specifically designed for the cleavage of proteins at aspartic acid, with the option of also performing online reduction by the introduction of dithiothreitol (DTT).[6] Hauser and Basile modified a standard CEM reaction vessel by drilling two threaded holes into the top cap and then fitted the system with two Rheodyne adaptors. A 1/16-inch tubing was connected directly to the port adaptor and a 5 μL microwave reaction loop was made from fused silica capillary. Each end of the reaction loop was inserted into the underside of the adapters. Water was added to the bottom of the reaction vessel and the top of the cell was tightened onto the bottom of the reaction vessel so that the microwave reaction loop seated completely inside the vessel. Figure 2.5a shows a schematic and Figure 2.5b a photograph of this online setup. As the actual microwave heating unit itself is not modified, this allows the system to remain operationally safe. Even after multiple uses, the integrity of the fittings and loops employed remained intact.

This system was also hooked up directly to a HPLC system which delivered a flow of 0.25 to 1 μL min^{-1}. For Asp-specific cleavage the following parameters were employed: 130 °C for 20 min with microwave power fluctuating between

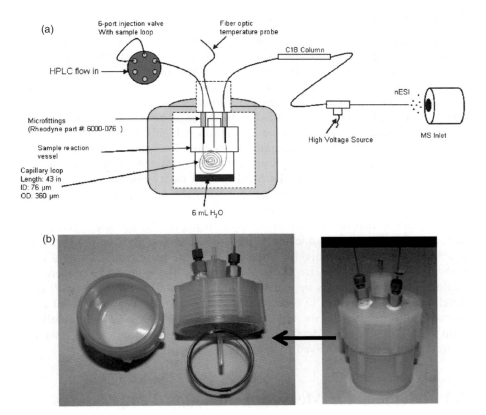

Figure 2.5 (a) Schematic of a microwave digestion flow cell for online microwave Asp-specific cleavage.[6] (Reproduced with permission from the American Chemical Society.) (b) Photograph of the online microwave Asp-specific cleavage apparatus. The flow cell is based upon a modification of the sample reaction vessel of a commercially available research-grade microwave oven. The reaction loop is 5 μL in total internal volume and is operated at a flow rate of 1 μL min^{-1} for a total digestion time of 5 min.[6] (Reproduced with permission from the American Chemical Society.)

50 and 250 W. The amount of time a sample was subjected to irradiation depended on the HPLC flow rate. Resultant proteolysis products from the Asp-specific cleavage could either be directly spotted onto a matrix-assisted laser desorption ionization (MALDI) plate for MALDI time-of-flight mass spectrometric analysis, or could be coupled to a reverse phase HPLC column for further separation and analysis by ESI-MS/MS. For increased sample coverage, online reduction could be performed whereby DTT was introduced with formic acid and in just a few minutes was shown to fully reduce the disulfide bonds of several test proteins. The potential of this type of setup is enormous and could potentially allow even further reductions in analytical protocols in the bottom-up analysis of proteins and protein mixtures.

2.2.3   High-Throughput Formats

Many commercial microwave systems designed for laboratory applications can be purchased with an autosampler which allows samples to be sequentially exposed to microwave-induced reactions. In proteomics, however, it is sometimes advantageous to prepare samples in a 96-well microtitre plate format, particularly for immunohistochemical techniques, or proteolysis experiments which are typically performed in a batch. In this case, being able to employ a 96-well plate format during microwave irradiation is highly beneficial and can be implemented as part of an automated workflow. For many years instrument manufacturers tried and failed to design a microwave system compatible with 96-well plates and finally in 2007 the MARS open cavity multimode microwave system from CEM was modified to accommodate this format. Indeed this system can be accommodated with a variety of turntables and vessel options with the 96-well plate format being optimal for high-throughput proteomic experiments. In this format the MARS microwave system can be modified with a turntable which can secure three individual 96-well microtitre plates. In addition a temperature probe can be inserted into one of the wells for accurate temperature readout and control. This system was employed by Zhu-Shimoni *et al.* who developed and compared two enzyme-linked immunosorbent assay (ELISA) formats for measuring the amount of Protein A leached from the immunoaffinity resin[7] (see Chapter 8 for details of their methodology). Figure 2.6 shows the 96-well plate format employed by the MARS microwave system.

2.3   Methods for Measuring Reaction Temperatures

The temperature of a microwave-assisted reaction can be measured by employing one of two different methods during microwave exposure: using either an infrared pyrometer, which detects the temperature of the surface layer, or a fluoroptic or fiber-optic temperature probe inside the vessel, which measures the temperature at a specific *in situ* position.[8] The most commonly reported mechanism of temperature measurement and control during proteomics experiments is the incorporation of an *in situ* fiber-optic temperature

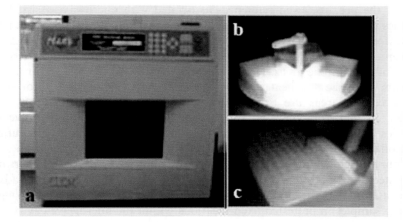

Figure 2.6 Photograph of the CEM MARS microwave instrument: (a) exterior image; (b, c) interior of the system along with the 96-well plate setup.[7] (Reproduced with permission from Judith Zhu-Shimoni, Genentech Inc.)

Table 2.1 Number of publications on microwave-assisted proteomics and the instrumentation employed.[9]

Year	Domestic microwave	Application microwave[a]
2002	3	2
2003	–	–
2004	2	3
2005	6	4
2006	2	3

[a]Systems cited include the CEM MDS 2100 model and CEM Discover model microwaves.

probe into the reaction media (typically by allocating a mock sample into which the probe can be inserted during the reaction). Using this configuration, one can ensure fast, accurate and convenient reaction feedback control. With several commercially available laboratory microwave instruments the probe can control temperature by either decreasing the magnetron power as necessary, or *via* the introduction of a cooling mechanism, such as a blast of air or a given gas (*e.g.* nitrogen).

2.4 Commercially Available Instrumentation

A summary of the number of publications employing microwave assisted proteomic reactions either in a domestic or a commercial application microwave was published by Lill *et al.*[9] (Table 2.1). This table demonstrates that over half of the peer reviewed references associated with microwave-assisted proteomics employ a domestic multimode open cavity microwave system such as those found in an everyday kitchen. Some modern cavity microwaves can deliver a very even field density, enabling microwave heating for use on a wide range of biocatalytic areas. However, these microwave systems are prone to a random dispersion of heat which can lead to the generation of "hot spots" where certain areas are exposed to significantly higher temperatures than others. In the proteomic and protein chemistry arena it is imperative that samples be treated in a homogeneous manner and temperatures be accurately controlled, especially when dealing with potentially heat-labile enzymes. To minimize the effect of uneven temperature distribution, several researchers have suggested adding beakers of cold water to domestic microwave ovens to adsorb excess thermal energy. In addition to this idea, some researchers also suggest placing samples in pretested fixed locations within the microwave cavity, with the aim of improving sample reproducibility. However, for more controlled and homogeneous reactions, application-specific microwave apparatus offers superior sample handling as compared to the use of domestic microwave ovens for biochemical reactions.

There are many systems currently marketed with specific biological and biochemical applications in mind. For example, CEM markets a system specifically for performing high-throughput tryptic digestions. This system is comprised of the Discover® system with a screw-top container capable of holding multiple microvials or Eppendorf tubes along with an insert for a fiber-optic temperature probe. In this setup, the fiber-optic probe can help stabilize the temperature by monitoring the magnetron power while inducing a simultaneous cooling function to allow energy input while maintaining the necessary relatively cool temperature required for such enzyme-driven proteolytic reactions.

In addition, several companies market apparatus with amino acid hydrolysis in mind (see Chapter 6 for a full description of microwave-assisted acid hydrolysis for amino acid analysis). Again, in conjunction with the Discover® microwave unit from CEM, a 45 mL vapor-phase hydrolysis vessel is available which allows for the processing of up to 10 (300 μL) samples in parallel. The system includes a valve panel which allows connection of the hydrolysis vessel to a vacuum and nitrogen source. The sealed sample vessel is alternately vacuum evacuated and purged with nitrogen allowing hydrolysis to be performed under inert, anaerobic conditions, therefore preventing oxidative degradation of the proteins/peptides. A schematic of the CEM microwave vapor-phase hydrolysis system is shown in Figure 2.7.

When considering investing in an application-specific microwave system one has to consider the technical and throughput benefits and limitations as compared to incorporation of a domestic microwave oven. For reactions carried out at high temperature (such as acid-induced hydrolysis or the Akabori reaction for C-terminal peptide sequence identification), or reactions not requiring a precise and even distribution of heat dissipation (for example during the staining and destaining of analytical gels), it may not be necessary to

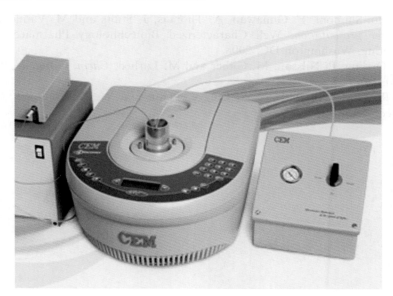

Figure 2.7 Image of the Discover® microwave unit from CEM with a 45 mL vapor-phase hydrolysis vessel allowing for the processing of up to 10 (300 μL) samples in parallel. The system includes a valve panel which allows connection of the hydrolysis vessel to a vacuum and nitrogen source. The sealed sample vessel is alternately vacuum evacuated and purged with nitrogen allowing hydrolysis to be performed under inert, anaerobic conditions, therefore preventing oxidative degradation of the proteins/peptides.

purchase a relatively expensive application microwave. However, for a high-throughput protein chemistry or proteomic laboratory employing proteolytic enzyme reactions on a daily basis, a commercial application microwave is worth its weight in gold.

References

1. *Electronic Design for Engineers and Engineering Managers*, Hayden Publishing, Rochelle Park, NJ, 1976, vol. 24.
2. T. J. Morgan, *RADAR*, F. Muller, London, 1960.
3. R. Buderi, *The Invention that Changed the World: How a Small Group of Radar Pioneers Won the Second World War and Launched a Technological Revolution*, Simon and Schuster, New York, 1996.
4. Developments in Microwave Chemistry, Intellectual Property Report, Evalueserve Analysis, 2005.
5. E. Comer and M. G. Organ, *J. Am. Chem. Soc.*, 2005, **127**, 8160.
6. N. J. Hauser and F. Basile, *J. Proteome Res.*, 2008, **7**, 1012.

7. J. Zhu-Shimoni, F. Gunawan, A. Thomas, J. Stults and M. Vanderlaan, *Poster presentation*, Well Characterized Biotechnology Pharmaceuticals meeting, Washington DC, 2008.
8. K. Orrling, P. Nilsson, M. Gilber and M. Larhed, *Chem. Commun.*, 2004, 790.
9. J. R. Lill, E. S. Ingle, P. S. Liu, V. Pham and W. N. Sandoval, *Mass Spectrom. Rev.*, 2007, **26**, 657.

CHAPTER 3

Mechanisms of Microwave-Assisted Action

Abstract

An ongoing discussion has been pursued by synthetic and medicinal chemists over the past few decades as to the exact mechanism of microwave catalysis. The main conundrum posed is: *does microwave irradiation purely catalyze reactions through heat, or through a combination of thermal and non-thermal energies?*

There are several hypotheses as to the mechanisms of action for microwave-assisted reactions, in organic synthesis, medicinal chemistry and more recently protein chemistry. This chapter explores the potential mechanisms involved and summarizes the outcomes as regards microwave applications in the world of proteomics and protein chemistry.

3.1 Dipolar Polarization Mechanisms

Before embarking on a discussion as to the mechanisms of action for microwave-assisted catalyses, one has to consider the following point. If two samples, one containing water and the other containing dioxane (or an equivalent non-polar solvent), are heated at equal power and length of time in a single-mode microwave cavity, the final temperature of the water will be significantly higher than that of the dioxane.[1] In order to fully understand this concept the following mechanisms have to be defined.

Microwave-assisted methods of catalysis in general involve the dielectric effect of microwave heating. As with all electromagnetic irradiation, microwave radiation can be divided into an electric field component and a magnetic field component.[1] It is the electric field component that is ultimately responsible for

Microwave-Assisted Proteomics
By Jennie Rebecca Lill
© Jennie Rebecca Lill 2009
Published by the Royal Society of Chemistry, www.rsc.org

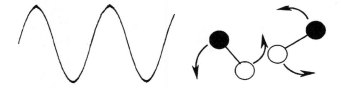

Figure 3.1   Dipolar molecules try to align with an oscillating electric field.[1] (Reproduced with permission from Elsevier Science and Technology Journals.)

the dielectric heating. The first component of dielectric heating is induced by what is termed the dipolar polarization mechanism.

In broader terms, microwaves can catalyze reactions by inducing molecular perturbation by a stimulation of ionic diffusion, but also by enhancement of dipole rotation without causing any rearrangement of molecular structures. In order for a substance to generate heat when irradiated with microwaves, it must possess a dipole moment (as is the case for water in the example above). A dipole is sensitive to external electric fields and will continually attempt to realign itself with the field by rotation, as demonstrated in Figure 3.1.[1] It is this applied field that provides energy for rotation of the molecule to occur.

This is how microwave-assisted heating differs from conventional heating: the dipole rotation constitutes an alternative efficient form of molecular agitation and it is due to this added molecular agitation that increased molecular catalysis is thought to occur.[2,3]

Polar molecules align themselves with the external applied field and heating occurs due to the torsional effect as polar molecules rotate back and forth, continually realigning themselves with the changing field (for a microwave irradiation frequency of 2450 MHz this equates to field changing 2.45×10^9 times per second). It is because of this dielectric heating and dipole polarization mechanisms that certain solvents such as hexane, toluene and dioxane are effectively transparent to microwaves, whereas water and methanol are easily heated by microwave irradiation.[4]

Collins and Leadbeater described how this mechanism can be translated into the microwave-assisted heating of protein and enzyme substrates.[5] The substructures of proteins, especially the secondary structures such as α-helices and β-sheets, contain high-density regions of hydrogen bonding which create a stack of peptide bond dipoles which are added across the hydrogen bonds within the protein. This leads to a theoretical large net dipole effect from one end of the secondary structure to the other. The dipole moments calculated for a number of proteins are summarized in Table 3.1.

The presence of a large net dipole moment across protein secondary structures, in particular α-helices, may add towards increased digestion rates of some proteins upon microwave irradiation. Figure 3.2 shows a theoretical mechanism of how microwave energy may interact with the dipole of the α-helix of a protein. Perturbation of the three-dimensional structure (*i.e.* denaturation) of the protein may result which could facilitate digestion of previously enclosed areas of the protein.

Table 3.1 Comparative dipole moments.[5] (Reproduced with permission from RSC Publishing.)

Molecule	Dipole moment (debye)
Water	1.8
Peptide bond	3.5
Myoglobin	170
Horse serum albumin	380
Horse carboxy hemoglobin	480

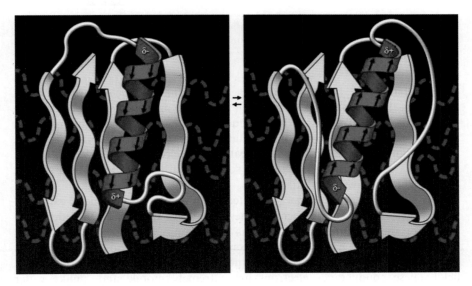

Figure 3.2 Proposed mechanism of action for a dipole moment across an α-helix and its interaction with microwave radiation.[5] (Reproduced with permission from RSC Publishing.)

3.2 Conduction Mechanism

A perfect scenario to explain the concept of microwave-induced conduction mechanisms is to compare the heating of two water samples, one containing distilled water and the other tap water, again in a single-mode microwave cavity at fixed radiation power and incubation time. The final temperature will be slightly higher in the tap water sample than in the distilled water sample.[1] Ions, or any substance with a hydrogen bonded cluster, will move through a solution under the influence of an electric field. This results in the expenditure of energy due to increased molecular collision rates and the resultant kinetic energy is

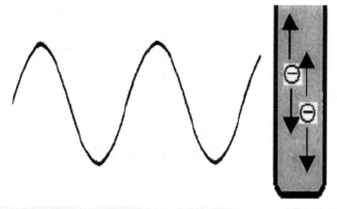

Figure 3.3 Charged particles in a solution will follow the applied electric field.[1]
(Reproduced with permission from Elsevier Science and Technology
Journals.)

converted to heat. The conduction mechanism has a much more profound
effect than the dipolar mechanisms with regards to microwave-assisted heating.
The heat generated through this conductive mechanism is accumulative with
the heat generated through dipole rotation and these two mechanisms often
work in tandem to cause increased heating of microwave-catalyzed solvents.[1]
Figure 3.3 demonstrates how ions in solution will follow the applied electric
field during microwave irradiation.

Another mode of conductance is related to the employment of metal-catalyzed
reactions, for example the chemiluminescence work pioneered by Geddes and
Aslan for the quantitation of proteins by immunoassay[6] (see Chapter 8 for
more detail). In the case of metals, microwave-assisted heating arises from the
creation of currents resulting from charge carriers being displaced by the electric
field. These conductance electrons are extremely mobile and, unlike water
molecules, can be completely polarized in 10–18 s. Although the common theory
is that "you shouldn't put metal in the microwave", arcing only occurs when
the metal particles are large, or are arranged in a continuous strip. If the
metal is organized in such a manner then large potential differences can result,
which lead to dramatic discharges if the potential differences are significant
enough to break down the electric resistance of the medium separating the
large metal particles. In the work by Aslan and Geddes, this phenomenon was
overcome by employing small metal particles which did not generate sufficiently
large potential differences for the "arcing" phenomenon to occur. Heating of the
surrounding matrix (*i.e.* the proteins, solvents (buffers) and fluorophore
reagents employed for visualization and quantitation of the immunoassay)
occurred due to the charge carriers being displaced by the microwave-
associated electric field which are subject to resistance in the medium in which
they travel. This leads to an overall ohmic heating of the metal nanoparticles in
addition to the heating of any surface polar molecules. This metal-associated

heating in combination with the microwave-assisted heating of the solvent leads to rapidly accelerated assay kinetics.[6]

3.3 Superheating Theory

There are multiple examples cited in the literature where microwave assistance is depicted as being nothing more than a rapid heating process. Stuerga *et al.* suggested that as a result of the more rapid heating that microwaves induce, the reaction time of chemical syntheses may be decreased by up to 75%. However, they suggest that this decrease in reaction time is purely due to superheating and not to any increased non-thermal molecular perturbation.[3,7] Many experts now believe that the microwave effect is the result of the microwaves' ability to superheat solvents beyond their normal boiling points. For example, water reaches 105 °C before boiling in a microwave oven, whereas acetonitrile boils at 120 °C instead of its usual 82 °C.[8] In a pure solvent, the higher boiling points achieved through superheating can be maintained but only while the microwave irradiation is being applied. However, the presence of any ions or contaminants in the solvent will contribute to the formation of what are termed "boiling nuclei" which aid the return of the overall solvent temperature to the "normal" boiling point.[1] Although this mechanism is typically more applicable to organic chemistry reactions, it may also hold true in some of the proteomic reactions, for example chemical cleavages and microwave-assisted acid hydrolysis reactions, which are performed at higher temperatures.

3.4 Mechanisms of Action Specific for Proteomic and Protein Chemistry Analyses

As stated above, the theory of superheating could be one explanation for the increased reaction rates observed for chemical cleavage of proteins and post-translational modifications, and indeed protein hydrolysis in the case of the Akabori reaction and microwave-assisted acid hydrolysis. However, for biological enzymes, superheating would not be an apt theory, as these molecules are very sensitive and are heat-labile. For example, Vesper *et al.* demonstrated that the heat-labile enzyme Glu-C has decreased proteolysis under microwave-assisted incubation, because the excess energy actually renders this fragile enzyme inert.[3,9] In the case of more resilient techniques such as the microwave-assisted acid hydrolysis methods for cleavage of the termini, or acid-labile sites within a protein, microwave irradiation may assist in protein denaturation and solubilization, therefore bringing the protein molecule into closer proximity with the solvent and reactive reagent trifluoroacetic acid, which contributes to its rapid hydrolysis.[3]

One has to be careful in discerning if a proteomic method has truly benefited from a microwave-mediated catalysis, or if the same results could be achieved through higher temperature non-microwave heating mechanisms. Indeed when scanning the literature one must be wary of identifying a true overall direct comparison between the microwave-mediated and convection-mediated catalysis. For example Sun *et al.* presented data on the microwave-assisted protein enzymatic digestion (MAPED) method.[10] In these MAPED experiments a comparison was performed between a standard protein preparation technique and a microwave-assisted approach. The overall conclusion was that preparation of a protein enzymatic digestion and extraction from a gel slice for characterization by mass spectrometry could be performed within 25 min with the MAPED technique as compared to the 19.5 h it takes to prepare the sample in a conventional manner. However, zero time-points were included for reduction and alkylation steps with the MAPED approach which suggested that these essential steps were omitted, yet 1 h and 30 min, respectively, were employed for the comparison to the standard protocol. If one were to perform the exact same comparison between both sample preparation methods, then it would still be true that the microwave-mediated method was the faster; however, Lill *et al.* found that reduction and alkylation on simplified samples of proteins could be 100% complete within 2 min either using a water bath or microwave on a simple protein set, and therefore, without undermining such work, one has to be careful about oversimplification of data presentation.[3]

There are, however, many examples in the literature where true direct comparisons between microwave-assisted catalysis and water bath-mediated incubations have been performed using identical temperatures, vessels and time points.[9,11–13] For example, Sandoval *et al.* described attempts to validate the non-thermal effects theory by heating vessels and buffers to the same temperature prior to commencement of incubation (*i.e.* microwave *vs.* water bath). This comparison was performed to determine whether microwave assistance merely "kick-started" the reaction by achieving optimum temperature faster than conventional heating methods. Although an overall increase in catalysis by both methods was observed by pre-heating, the explanation as to why a one-hour incubation of the glycoprotein RNase-B with PNGase-F leads to complete deglycosylation at 37 °C in the microwave, but at the same temperature with an overnight incubation in the water bath *N*-linked deglycosylation is still not complete, cannot explain the theory with thermal events alone. This experiment provided some evidence that non-thermal microwave effects may contribute to the catalysis observed using microwave irradiation.[13]

So with all this in mind, does microwave assistance aid in increasing the kinetic rates for all protein chemistries? Pramanik *et al.*[11] stated that "No microwave-assisted reaction is slower than the corresponding conductivity/convection reaction". However, although this statement may be true for the majority of chemical cleavages of proteins and posttranslational modifications, it is not necessarily true for heat-labile proteolytic enzymes such as Glu-C.[3,9] When considering the use of microwave assistance to increase throughput, each enzymatic or chemical preparation should be fully investigated on a case-by-case basis. While an apparent rate enhancement has been reported for multiple microwave-assisted proteomic and protein chemistry experiments, a larger data set is required before any definite conclusions can be made about the exact mechanisms involved.

References

1. P. Lidstrom, J. Tierney, B. Wathey and J. Westman, *Tetrahedron*, 2001, **57**, 9925.
2. C. Strauss and R. Trainor, *Aust. J. Chem.*, 1995, **48**, 1665.
3. J. R. Lill, E. S. Ingle, P. S. Liu, V. C. Pham and W. N. Sandoval, *Mass Spectrom. Rev.*, 2007, **26**, 657.
4. M. J. Previte, K. Aslan and C. D. Geddes, *Anal. Chem.*, 2007, **79**, 7042.
5. J. M. Collins and N. E. Leadbeater, *Org. Biomol. Chem.*, 2007, **5**, 1141.
6. K. Aslan and C. D. Geddes, *Anal. Chem.*, 2005, **77**, 8057.
7. D. Stuerga, K. Gonon and M. Lallemant, *Tetrahedron*, 1993, **49**, 6229.
8. D. Adam, *Nature*, 2003, **421**, 571.
9. H. W. Vesper, L. Mi, A. Enada and G. L. Myers, *Rapid Commun. Mass Spectrom.*, 2005, **19**, 2865.
10. W. Sun, S. Gao, L. Wang, Y. Chen, S. Wu, X. Wang, D. Zheng and Y. Gao, *Mol. Cell Proteomics*, 2006, **5**, 769.
11. B. N. Pramanik, U. A. Mirza, Y.-H. Ing, Y.-H. Liu, P. L. Bartner, P. C. Weber and A. K. Bose, *Protein Sci.*, 2002, **11**, 2676.
12. S. S. Lin, C.-H. Wu, M.-C. Sun, C.-M. Sun and Y.-P. Ho, *J. Am. Soc. Mass Spectrom.*, 2005, **16**, 581.
13. W. N. Sandoval, F. Arellano, D. Arnott, H. Raab, R. Vandlen and J. R. Lill, *Int. J. Mass Spectrom.*, 2007, **259**, 117.

CHAPTER 4
Microwave-Assisted Enzymatic Digestions

Abstract

Despite advances in "top-down" mass spectrometric methods for characterizing proteins, the traditional "bottom-up" approach of digesting a protein into smaller peptides followed by either peptide mass fingerprinting or reverse-phase separation and tandem mass spectrometric identification remains the most widely employed analytical method for protein characterization to date. Proteolytic digestion remains a rate-limiting step in sample preparation, and many traditional protocols recommend overnight incubation of the substrate and enzyme at physiological temperature. However, it was demonstrated in several recent studies that many proteolytic enzymes can tolerate temperatures higher than 37 °C. In addition, it was shown that non-conventional incubation methods such as ultrasonic vibration or microwave irradiation can also accelerate these typically lengthy reactions. This chapter summarizes the utility of microwave-assisted enzymatic digestion for "bottom-up" proteomic approaches and explores associated innovative sample preparation methods that have further accelerated these proteolytic digestions.

4.1 Introduction to Proteomic Work Flows

Advances in genomic and proteomic technologies have resulted in the production of many bio-therapeutics composed of recombinant proteins.[1] Mass spectrometry is typically the primary tool employed for the initial identification and subsequent full characterization of proteins and their posttranslational modifications (PTMs). Characterization of either a single protein or a complex mixture of proteins using mass spectrometry typically involves either (a)

Microwave-Assisted Proteomics
By Jennie Rebecca Lill
© Jennie Rebecca Lill 2009
Published by the Royal Society of Chemistry, www.rsc.org

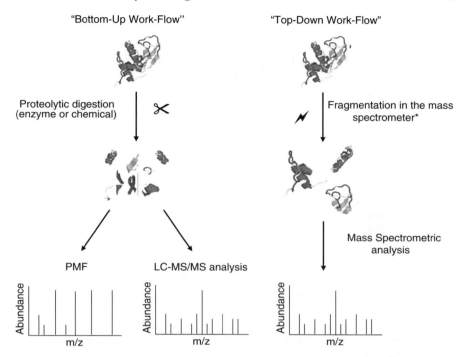

Figure 4.1 An overview of "bottom-up" and "top-down" proteomic work flows for the characterization of proteins.

digestion of the protein(s) with a proteolytic enzyme or chemical followed by mass spectrometric analysis, a protocol typically referred to as the "bottom-up" approach,[2] or (b) fragmentation of the intact protein in the mass analyzer of the spectrometer with no prior proteolytic digestion *via* a high-energy dissociation method, a process typically referred to as the "top-down" approach.[3,4] Figure 4.1 summarizes both of these approaches for the mass spectrometric characterization of proteins.

Bottom-up proteomics can also be further sub-categorized into two types of analysis: peptide mass fingerprinting (PMF)[5] or LC-MS/MS.[6] In PMF, peptides are analyzed in full (*i.e.* no fragmentation), mainly by matrix-assisted laser desorption ionization (MALDI) time-of-flight (TOF) mass spectrometric analysis. Masses corresponding to the intact molecular weight of each peptide can be pieced together to identify a unique protein fingerprint and spectra are either interpreted manually or through informatic processing with a search algorithm such as Mascot (Matrix Science, London, UK).[7] Alternatively, LC-MS/MS can be performed, whereby peptides are separated by reverse-phase chromatography which is typically coupled online to a mass spectrometer. Here, peptides are ionized as they are eluted from the chromatography column with electrospray or nanospray ionization and subsequently

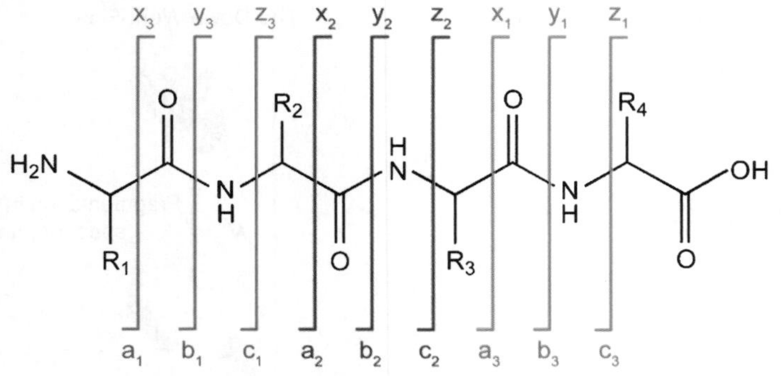

Figure 4.2 Nomenclature for peptide fragment ions. Fragment ions with the charge retained at the N-terminus of the peptide are denoted b-ion series and those at the C-terminus of the peptide as y-ion series.[8]

fragmented using tandem mass spectrometry (MS/MS). For this method, fragmentation can be induced by a number of processes; however, the most common fragmentation mechanism for general peptide identification occurs when peptide ions collide with an inert gas (*e.g.* helium) under a high voltage in a process typically referred to as collision-induced dissociation (CID). During CID peptides fragment along their peptide backbone in the mass analyzer and a series of characteristic fragment ions are formed. In CID an ion produced from the cleavage of the peptide at the peptide bond with a charge retained at the N-terminus is referred to as a b-ion and that at the C-terminus as a y-ion (see Figure 4.2).[8] The sequence of the peptide that has undergone fragmentation can be interpreted by calculating the mass differences between the b-ion series and the converse y-ion series. Many search algorithms such as Sequest,[9] Mascot[7] and Inspect[10] can easily process this data and are employed to identify peptide and therefore protein sequences in an automated manner.

4.2 Microwave-Assisted Tryptic Digestions

Trypsin is a 24 kDa protein belonging to the serine hydrolase enzyme family and for commercial use is typically derived from porcine pancreas. Trypsin is perhaps the most commonly employed proteolytic enzyme in "bottom-up" proteomics as it converts most protein mixtures into more readily analyzable peptide populations. Trypsin cleaves at the carboxyl termini of arginine (Arg) and lysine (Lys) (except on occasion when sterically hindered by the presence of a neighboring proline residue). These two amino acids have an ideal prevalence in the majority of proteins for yielding optimal sized peptides for analysis by LC-MS/MS. In addition, due to the basic nature of these two amino acids, tryptic peptides typically ionize well and produce good fragmentation patterns by most tandem mass spectrometric protocols due to their ability to carry a charge at the C-terminus of the peptide.[11]

Tryptic digestion is conventionally carried out at physiological temperature (37 °C) in a water bath or convector oven for lengthy periods (8 h to overnight) for complete hydrolysis. As a result, researchers have investigated and adopted many protocols to decrease proteolytic digestion times. Such protocols have included adding a small percentage of organic solvents such as acetonitrile or methanol, detergent (urea) or acid-labile surfactants (RapiGest™) to digestion buffers to increase the digestion efficiencies and decrease incubation times, typically by enabling further denaturation of the protein for better exposure of cleavage sites to the proteolytic enzyme.[12–14] Immobilization protocols have also been explored to further accelerate proteolysis, for example by introduction of immobilized enzymes to solid supports[15,16] or *via* immobilization of proteins to poly(vinylidine difluoride) (PVDF) membranes followed by incubation with non-ionic surfactants and proteolytic enzymes.[17] More recently, methods employing alternative energies such as ultrasonic vibrations (*via* a sonoreactor or ultrasonic probe)[18] or microwave irradiation have been employed to increase proteolytic catalysis.[19,20] Over the past seven years or so, microwave-assisted tryptic proteolysis has been described in multiple studies and adopted as a standard protocol in many laboratories.[21]

Accelerated proteolytic cleavage of proteins under controlled microwave conditions (*i.e.* set temperature, pressure and power) in a scientific monomode microwave system was first described by Pramanik *et al.*[19] It was demonstrated that microwave-assisted digestions could be achieved using the proteolytic enzymes endoproteinase Lys-C (Lys-C) and trypsin. MALDI-TOF analysis using a Voyager-DE STR mass spectrometer allowed detection of both the intact protein (start material) and resultant proteolysis products by PMF. Sequence coverage (*i.e.* how much of the protein sequence was identified *via* characterization of the proteolysis products) and quantitation were performed by LC-MS/MS using a PE-Sciex API 365 triple quadrupole mass spectrometer. This facilitated assessment of the rate of completion of proteolytic cleavage allowing a direct comparison between the microwave-assisted method and traditional tryptic digestion techniques. The first study conducted was performed on bovine cytochrome C, a globular protein known to be relatively resistant to traditional proteolytic cleavage methods. After a mere 10 min microwave exposure at 37 °C, peptides covering a large percentage of the protein could be detected by MALDI-TOF mass spectral analysis. Complete digestion of several non-reduced, tightly folded proteins could be completed in 12 min in the presence of microwave radiation; however, in a water bath at an identical time point proteolysis was not observed.[19] Figure 4.3 shows the MALDI-TOF mass spectra of cytochrome C after a 12 min tryptic digestion (1:25 enzyme-to-substrate ratio) with microwave assistance and conventional water bath incubation, at identical temperatures (37 °C).

The level of proteolysis achieved after a 12 min tryptic digestion of cytochrome C using microwaves at 37 °C was comparable to that observed for a 6 h incubation at the same temperature in a water bath. Additional proteins were employed for evaluating microwave-assisted tryptic digestion, and in addition to demonstrating that microwave-assisted conditions could be applied

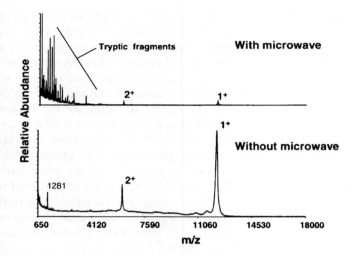

Figure 4.3   MALDI mass spectra of cytochrome C after a 12 min tryptic digestion
(1:25 enzyme-to-substrate ratio) with microwave assistance and conventional water bath incubation at identical temperatures (37 °C).[19] (Reproduced with permission from *Protein Science*.)

universally for accelerated in-solution tryptic digestions, Pramanik *et al.* also
showed that these protocols could be successfully applied to in-gel tryptic
digestions, which was interesting in that proteins embedded in a semi-immobile
matrix could also benefit from microwave assistance. Destaining of the gel prior
to enzymatic incubation (as described in Chapter 2) was also performed in the
microwave system and in-gel microwave-assisted digestions over 15 min could
provide high peptide recovery and protein coverage for bottom-up protein
characterization.

Several key observations were made from this initial study. Firstly, when the
enzyme was omitted from the microwave-assisted incubation, the substrate
protein remained intact and was not cleaved by heat-induced proteolysis or
self-induced degradation, demonstrating that accelerated proteolysis was not
random. Secondly, sites of proteolysis remained specific, and even when samples were incubated with microwave assistance at 60 °C, cleavage occurred
specifically at predicted Lys and Arg residues for tryptic cleavage. Further-
more, the observed kinetics varied significantly between water bath and
microwave-assisted incubations. By sampling aliquots over time periods from
5 to 30 min, it was concluded that, in contrast to the water bath incubation,
microwaves accelerated proteolysis in the first few minutes of incubation,
diminishing rapidly after 30 min, suggesting that the enzyme was denatured and
inactivated in a short period of time upon microwave exposure. Figure 4.4 plots
the various temperatures investigated for both microwave-assisted and water
bath incubated tryptic digestions of IFN-2b.[19]

No artifactual effects such as significantly higher deamidation or oxidation
were observed as a direct result of microwave-assisted incubations. Overall,

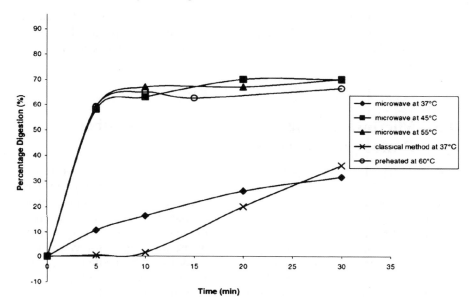

Figure 4.4 Plots showing the rate of trypsin digestion of IFN-α-2b in the first 30 min under microwave irradiation and classic conditions.[19] (Reproduced with permission from *Protein Science*.)

Pramanik *et al.* showed that 60 °C is the optimum temperature for proteolysis, that proteolysis is greatly enhanced when mediated by microwave radiation and that tightly folded proteins, typically requiring hours of incubation under conventional methods, benefit the most from this microwave-assisted proteolysis.[19]

A quantitative assessment of microwave-assisted enzymatic proteolysis with enzymes trypsin and endoproteinase Glu-C (Glu-C) was performed on glycated hemoglobin A1c by Vesper *et al.*[20] An LCQ Deca ion trap mass spectrometer was used for qualitative analysis and a TSQ Quantum triple quadrupole mass spectrometer for quantitation of peptides corresponding to tryptic and Glu-C digestion products from the N-terminus of the glycated hemoglobin A1c protein. For tryptic digestions, complete digestion was obtained at 50 °C within 20 min; however, at temperatures above 55 °C the efficiency of digestion substantially decreased, probably due to heat-inactivation of the enzyme.[20] This optimum microwave-assisted temperature therefore varied by 10 °C from the optimal temperature described by Pramanik *et al.*, although it was still proved that tryptic digestions could be successfully performed at higher temperatures using microwaves than the conventionally cited 37 °C.

Vesper *et al.* performed an investigation assessing microwave-assisted tryptic digestions reproducibility compared to traditional water bath mediated digestions. By comparing the traditional tryptic digestion protocol to a microwave-assisted method, close correlation between the two methods was observed,

demonstrating that microwave-assisted enzymatic digestion provided quantitative results similar to traditional water bath digestion protocols. For in-
solution and in-gel microwave-assisted protocols, refer to protocols III and IV
in Chapter 10.

4.3   Microwave-Assisted Proteolysis with Other Proteolytic Enzymes

Although trypsin is the most commonly employed enzyme for bottom-up
proteomic analysis, a plethora of other enzymes are utilized for more in-depth
analyses; for example, to ensure that complete coverage of a protein is achieved
for *de novo* sequencing projects,[22,23] for the characterization of posttranslational modifications or for protein isoform determination. Also, with the
increased popularity of employing top-down methodologies for larger polypeptides, in a technique often referred to as "middle-down" proteomics, Lys C,
Glu-C, endoproteinase Asp-N (Asp-N) and other enzymes producing larger
(15–50 amino acids) peptide fragments are commonly employed.[24] Vesper *et al.*
examined the potential of microwave-mediated Glu-C proteolytic digestions;
however, these were shown to yield fewer proteolytic products than the conventional convection heating method. Autolytic peaks from the Glu-C enzyme
were not detected; therefore it was concluded that inactivation of the enzyme
was due to microwave-induced denaturation resulting from the instability of
this enzyme at elevated temperatures and not due to autolysis.[20]

Lill *et al.* conducted experiments to determine whether microwave-assisted
digestions can increase proteolysis for several enzymes for mass spectrometric
identification.[25] Bovine serum albumin (BSA) and myoglobin (50 pmol) were
subjected to either (a) tryptic digestion (0.1 µg trypsin in 25 mM ammonium
bicarbonate, pH = 7.5) in a water bath at 37 °C or using a CEM Discover
microwave system at 37 °C with 2–5 W of power applied, or (b) Asp-N digestion
(0.5 µg Asp-N in 50 mM sodium phosphate, pH = 7) in the microwave system or
water bath. Incubations proceeded for 5, 10 or 30 min, and were stopped by the
addition of 0.1% trifluoroacetic acid followed by storage at −20 °C. Samples
were analyzed on a sodium dodecyl sulfate polyacrylamide gel electrophoresis
(SDS-PAGE) gel and visualized by staining with Coomassie blue as this method
allowed simple semi-quantitative and qualitative visualization for comparing
results. Results for these time points are shown in Figure 4.5a for tryptic
digestion and Figure 4.5b for Asp-N digestion. Microwave-assisted tryptic
digestion occurred at a much faster rate than the water bath incubated digestion.
After 1 h, the microwave-assisted tryptic proteolysis was complete (or after
30 min at a higher temperature of 60 °C with the microwave system; data not
shown). Microwave-assisted Asp-N digestions did not show any significant
reduction in proteolysis time over conventional water bath incubations; however, proteolysis was not decreased, as was previously observed for the heat-
labile enzyme Glu-C.[25]

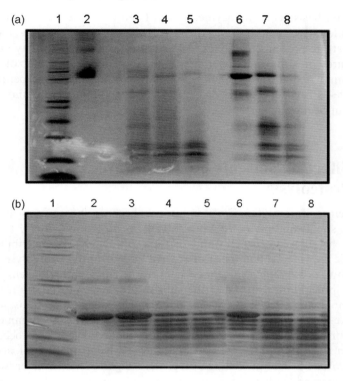

Figure 4.5 (a) SDS-PAGE analysis of BSA after tryptic digestion at 37 °C in both the water bath and the microwave oven. Lane 1, mark 12 molecular weight standard; 2, start material at 66 kDa; 3, microwave-assisted digestion 5 min; 4, microwave-assisted digestion 10 min; 5, microwave-assisted digestion 30 min; 6, water bath digestion 5 min; 7, water bath digestion 10 min; 8, water bath digestion 30 min. (b) SDS-PAGE analysis of myoglobin after Asp-N digestion at 37 °C in both the water bath and the microwave oven. Lane 1, mark 12 molecular weight standard; 2, start material at 66 kDa; 3, microwave-assisted digestion 5 min; 4, microwave-assisted digestion 30 min; 5, microwave-assisted digestion 1 h; 6, water bath digestion 5 min; 7, water bath digestion 30 min; 8, water bath digestion 1 h. (Reproduced with permission from *Mass Spectrometry Reviews*.[25])

Lys-C is the most commonly employed enzyme in middle-down proteomics as, like trypsin, it cleaves leaving a basic residue at the C-terminus which can promote increased ionization and informative fragmentation patterns. Lys-C microwave-assisted digestions were also investigated by Pramanik *et al.*, and this enzyme was shown to exhibit identical behavior to trypsin in that microwave-assisted Lys-C digestions could be performed in the presence of microwave radiation at elevated temperatures to rapidly produce proteolytic fragments for mass spectrometric analysis.[19]

Optimization of microwave conditions and a direct comparison to non-microwave-mediated proteolysis methods should be performed on a case-by-case basis, as each enzyme, even if from the same enzyme family, may behave differently. In addition to the heat lability of some enzymes, *e.g.* Glu-C, some substrate proteins or PTMs may be susceptible to instability at elevated temperatures or vibrational energies. This difference in behavior of substrate proteins may explain the slight differences in optimum temperatures for microwave-assisted enzymatic digestions cited in the literature and is further explored in Section 4.7.

4.4　Effect of Solvents on Microwave-Assisted Proteolysis

To accelerate proteolytic digestion, it has become increasingly popular to include a small amount of organic solvent in digestion buffers, to partially denature the substrate protein and therefore allow easier accessibility to the proteolytic enzyme. The amount of solvent added during proteolysis typically remains low, as too high a solvent concentration can denature the enzyme itself, or induce precipitation of either the substrate or enzyme. To see if this same solvent enhancing effect was true for microwave-assisted digestions, a study of microwave-assisted tryptic cleavages in the presence of various organic solvents was performed.[26] Enhanced tryptic digestions could indeed occur under microwave assistance in the presence of organic solvents (methanol, acetonitrile and chloroform). The percentage of protein digested under microwave irradiation increased with increasing amounts of acetonitrile, but decreased as methanol content was increased. Table 4.1 shows the sequence coverages of protein digestion for various solvent systems with and without microwave irradiation.

Increased rates of protein digestion in the presence of organic solvents were attributed to the denaturation of the protein and to differences in reaction temperatures employed with various solvent systems. Temperatures increased significantly faster in the presence of organic solvents as when compared to

Table 4.1　Sequence coverages (%) of protein digestions in various solvent systems with and without microwave irradiation.[26] (Reproduced with permission from *J. Am. Soc. Mass Spectrom.*)

Protein	H_2O	*50% CH$_3$OH*	*30% CH$_3$CN*	*49% CH$_3$OH/49% CHCl$_3$/2% H$_2$O*
Myoglobin	100 (96)	94 (100)	94 (100)	29 (0)
Cytochrome C	96 (100)	95 (15)	70 (14)	39 (0)
Lysozyme	36 (19)	21 (6)	30 (7)	20 (4)
Ubiquitin	42 (37)	80 (15)	53 (29)	20 (20)

The digestion efficiencies without microwave irradiation are indicated in parentheses. Digestion without microwave heating proceeded for 6 h at 37 °C. All the reactions under microwave irradiation, except for those in the experiments that involved CHCl$_3$, proceeded for 10 min at 60 °C. The experiments that involved CHCl$_3$ proceeded for 10 min at 50 °C.

purely aqueous solvents, hence initiating even further accelerated catalysis of the microwave-assisted tryptic digestion. These findings suggested that proteolytic kinetics were typically accelerated in the presence of organic solvents during microwave-assisted protein digestion. A further interesting observation was that, under rapid microwave irradiation, trypsin was still active even in the presence of chloroform; however, under traditional heating conditions the presence of chloroform rendered the enzyme inactive.[26]

4.5 Microwave-Assisted Protein Preparation and Enzymatic Digestion (MAPED) in Proteomics

In Chapter 1 several techniques were described whereby sample preparation, for example SDS-PAGE staining and destaining, could be accelerated in the presence of microwave radiation. Sun *et al.* explored the hypothesis of microwave-assisted protein preparation and coupled it to microwave-assisted tryptic digestion to coin the phrase "microwave-assisted protein preparation and enzymatic digestion in proteomics protocol" (MAPED). Here, each preparation step of a typical in-gel tryptic digestion for mass spectrometric analysis was performed with microwave assistance (Figure 4.6).[27]

Traditional sample preparation time was reduced from 19.5 h to 25 min when performed with microwave assistance as compared to conventional room temperature and water bath mediated incubations (Table 4.2). A conventional domestic microwave oven was employed for each microwave-assisted protocol and resultant peptide characterization was performed using an LCQ Deca XP Plus ion trap mass spectrometer (Thermo, San Jose, CA) and a Voyager DE-STR Pro MALDI-TOF mass spectrometer (Applied Biosystems, Foster City, CA). The MAPED approach was assessed with a single protein (BSA) in addition to complex protein mixtures (total yeast lysate and human urinary proteins) to test the applicability for both single-protein characterization and large-scale proteomic analysis. By analyzing highly complex samples with MAPED this approach was proven to be applicable to samples with a wide dynamic range and the MAPED protocol could be employed to produce a significantly higher number of identified and characterized MS/MS spectra than conventional incubation protocols in bottom-up analyses.[27]

4.6 Microwave-Assisted Tryptic Digestion on Immobilized Surfaces

In 1988 Walkeiwicz *et al.* published an article demonstrating that out of 150 substances examined, magnetite beads had optimal properties for microwave absorption.[28] Chen and Chen expanded on this concept and employed magnetite beads for accelerated microwave-assisted enzymatic digestions.[29] These multifunctional magnetite beads accelerated microwave-assisted digestions by

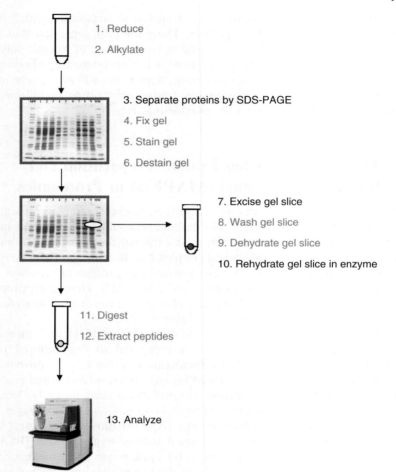

1. Reduce
2. Alkylate

3. Separate proteins by SDS-PAGE
4. Fix gel
5. Stain gel
6. Destain gel

7. Excise gel slice
8. Wash gel slice
9. Dehydrate gel slice
10. Rehydrate gel slice in enzyme

11. Digest
12. Extract peptides

13. Analyze

Figure 4.6   Typical protocol for the analysis of proteins using MAPED. Steps that can be drastically reduced by the introduction of microwave-assistance are highlighted in red.

their ability to absorb microwave radiation more efficiently than conventional solution-based proteolysis. Beads acted as "trapping probes" whereby the negatively charged functionality of the beads allowed proteins to adsorb to the surface of the beads due to electrostatic attraction, hence inducing an increased surface area of the protein, leading to a concentration effect of the protein near to the microwave-sensitive material. In addition, proteins became denatured and therefore more vulnerable to proteolysis once adsorbed to the beads surface.[29]

Digestions using microwave-assisted proteolysis on magnetite beads (150–600 μg) were shown to be dramatically increased, with complete proteolysis observed in as little as 30 s. Magnetite beads were multifunctional as they could be employed for microwave-assisted proteolysis by immobilization of

Table 4.2 Standard and MAPED in-gel preparation, digestion and peptide extraction protocols.[27] (Reproduced with permission from the American Society for Biochemistry and Molecular Biology.)

Protocol	Solution	Standard, sample 1 (h)	MAPED (min)			
			Sample 2	Sample 3	Sample 4	Sample 5
Reduce	10 nM dithiothreitol	1	60	60	5	0
Alkylate	55 mM iodoacetamide	0.5	30	0	0	0
Digest	0.5 µM trypsin, 25 mM ammonium bicarbonate	16	1	1	1	1
Total time		17.5	91	61	6	1

trypsin to the bead surface, rather than absorption of the protein for digestion. In addition the effect of solvent on digestion efficiencies was investigated and found (as described in Section 4.4) to further improve proteolysis when compared to aqueous-mediated magnetite bead proteolysis.[29]

In 2008 two articles were published by Lin *et al.* whereby novel microwave-assisted digestions were performed by trypsin immobilized on either magnetic silica microspheres[30] or nanoparticles.[31] These two methods followed the same concept as described by Chen and Chen;[29] however, in the first approach, trypsin was immobilized onto magnetic silica microspheres through a one-step reaction with 3-glycidoxypropyltrimethoxysilane (GLYMO) which provided the epoxy group as a reactive spacer. Using BSA and myoglobin as model proteins, it was demonstrated that peptide fragments from these proteins could be produced in as little as 15 s which were identified by MALDI-TOF.[30] In the second study, trypsin-immobilized magnetic nanoparticles (TIMNS) were produced and shown to be excellent microwave absorbers and hence ideal tools for microwave-assisted tryptic digestions.[31] In this study, as well as demonstrating the utility of the method for individual proteins, also demonstrated was the efficiency of the protocol to a protein extract from rat liver. Without any preparation and pre-fractionation, the entire proteome was digested by TIMNS in 15 s, which after LC-MS/MS analysis revealed 313 proteins (with $p < 0.01$). This demonstrated how immobilized trypsin microwave-mediated methods could be applied to large-scale high-throughput proteomic analysis.

4.7 Theory of Microwave-Assisted Proteolytic Digestions

Although increased incubation temperatures can equate to increased proteolytic digestion of proteins for several enzymes, it has been demonstrated by several

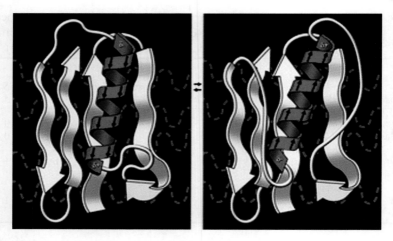

Figure 4.7   Dipole movement across an α-helix and interaction with microwave radiation.[30] (Reproduced with permission the American Chemical Society)

groups[19,21,25] that increased temperature alone is not the sole microwave-assisting parameter initiating increased catalysis. Collins and Leadbeater hypothesized that microwave-mediated effects on proteolysis could be due to increased dipole movements of the α-helices of proteins.[32] Most proteins possess α-helices and β-sheets as part of their tertiary structure and in α-helices the backbone of the polypeptide is coiled around the protein axis so that the side chains of the amino acids are pointing outwards and downwards from the backbone. This helical structure leads to a stack of peptide bond dipoles due to the large amount of hydrogen bonding within the protein structure (see Figure 4.7).

In turn this hydrogen bonding leads to an overall large net dipole effect across α-helices which, being susceptible to vibrational and structural rearrangements, in the presence of microwave radiation may lead to increased catalysis. If microwave energy induces perturbation of the three-dimensional structure of the protein, digestion could be facilitated by exposure of previously enclosed or buried regions of the protein to the proteolytic enzyme. This may explain why several groups have seen more dramatic effects of microwave-assisted proteolysis on non-denatured or reduced proteins with complex tertiary structures than studies performed on less structurally complex proteins.

In conclusion, microwave-assisted enzymatic reactions can drastically speed up proteolysis if both the substrate protein(s) and the enzyme are not heat labile and can serve as an invaluable tool for high-throughput bottom-up protein characterization.

References

1. Pharmaprojects (database online), Richmond, UK: PJB Online Services, 1980-.m. Updated weekly. Available on DataStar, Dialog, Ovid and STN.

2. J. Rappsilber and M. Mann, *Trends Biochem. Sci.*, 2002, **27**, 74.
3. R. Bakhtiar and Z. Guan, *Biotechnol. Lett.*, 2006, **28**, 1047.
4. G. C. McAlister, D. Phanstiel, D. M. Good, W. T. Berggren and J. J. Coon, *Anal. Chem.*, 2007, **79**, 3525.
5. W. J. Henzel, C. Watanabe and J. T. Stults, *J. Am. Soc. Mass Spectrom.*, 2003, **14**, 931.
6. A. L. McCormack, D. M. Schieltz, B. Goode, S. Yang, G. Barnes, D. Drubin and J. R. Yates 3rd, *Anal. Chem.*, 1997, **69**, 767.
7. D. N. Perkins, D. Pappin, D. M. Creasy and J. S. Cottrell, *Electrophoresis*, 1999, **20**, 3551.
8. P. Roepstorff and J. Fohlman, *Biomed. Mass Spectrom.*, 1984, **11**, 601.
9. J. R. Yates 3rd, J. K. Eng, A. L. McCormack and D. Schieltz, *Anal. Chem.*, 1995, **67**, 1426.
10. S. Tanner, H. Shu, A. Frank, L. C. Wang, E. Zandi, M. Mumby, P. A. Pevzner and V. Bafna, *Anal. Chem.*, 2005, **77**, 4626.
11. J. V. Olsen, S. E. Ong and M. Mann, *Mol. Cell Proteomics*, 2004, **3**, 608.
12. W. K. Russell, Z. Y. Park and D. H. Russell, *Anal. Chem.*, 2001, **73**, 2682.
13. Y. Q. Yu, M. Gilar, P. J. Lee, E. S. P. Bouvier and J. C. Gebler, *Anal. Chem.*, 2003, **75**, 6023.
14. A. Umar, J. C. Dalebout, A. M. Timmermans, J. A. Foekens and T. M. Luider, *Proteomics*, 2005, **5**, 2680.
15. J. Duan, Z. Liang, C. Yang, J. Zhang, L. Zhang, W. Zhang and Y. Zhang, *Proteomics*, 2006, **6**, 412.
16. G. Massolini and E. Calleri, *J. Sep. Sci.*, 2005, **28**, 7.
17. V. C. Pham, W. J. Henzel and J. R. Lill, *Electrophoresis*, 2005, **26**, 4243.
18. R. Rial-Otero, R. J. Carreira, F. M. Cordeiro, A. J. Moro, H. M. Santos, G. Vale, I. Moura and J. L. Capelo, *J. Chromatogr. A.*, 2007, **1166**, 101.
19. B. N. Pramanik, U. A. Mirza, Y. -H. Ing, Y. H. Liu, P. L. Bartner, P. C. Weber and A. K. Bose, *Protein Sci.*, 2002, **11**, 2676.
20. H. W. Vesper, L. Mi, A. Enadaand and G. L. Myers, *Rapid Commun. Mass Spectrom.*, 2005, **19**, 2865.
21. N. Wang, L. MacKenzie, A. G. De Souza, H. Zhong, G. Goss and L. Li, *J. Proteome Res.*, 2007, **6**, 263.
22. V. Pham, W. J. Henzel, D. Arnott, S. Hymowitz, W. N. Sandoval, B. T. Truong, H. Lowman and J. R. Lill, *Anal. Biochem.*, 2006, **352**, 77.
23. N. Bandeira, K. R. Clauser and P. A. Pevzner, *Mol. Cell Proteomics*, 2007, **6**, 123.
24. N. C. VerBerkmoes, J. L. Bundy, L. Hauser, K. G. Asano, J. Razumovskaya, F. Larimer, R. L. Hettich and J. L. Stephenson Jr, *J. Proteome Res.*, 2002, **1**, 239.
25. J. R. Lill, E. S. Ingle, P. S. Liu, V. Pham and W. N. Sandoval, *Mass Spectrom. Rev.*, 2007, **26**, 657.
26. S. S. Lin, C. -H. Wu, M. -C. Sun, C. -M. Sun and Y. -P. Ho, *J. Am. Soc. Mass Spectrom.*, 2005, **16**, 581.
27. W. Sun, S. Gao, L. Wang, Y. Chen, S. Wu, X. Wang, D. Zheng and Y. Gao, *Mol. Cell Proteomics*, 2006, **5**, 769.

28. J. W. Walkeiwicz, A. E. Clark and S. L. Mcgill, *Miner. Metall. Proc.*, 1988, **124**, 247.
29. W.-Y. Chen and Y.-C. Chen, *Anal. Chem.*, 2007, **79**, 2394.
30. S. Lin, G. Yao, G. Dawei, Y. Li, C. Deng, P. Yang and X. Zhang, *Anal. Chem.*, 2008, **80**, 3655.
31. S. Lin, D. Yun, D. Qi, C. Deng, Y. Li and X. Zhang, *J. Proteome Res.*, 2008, **7**, 1297.
32. J. M Collins and N. E. Leadbeater, *Org. Biomol. Chem.*, 2007, **5**, 1141.

CHAPTER 5

Microwave-Assisted Chemical Digestion of Proteins

Abstract

Chemical proteolysis is often employed as an alternative or complementary analytical tool to enzyme-mediated proteolysis for the characterization of proteins by mass spectrometry. Acids and other chemicals can be selected to cut at either specific amino acid residues or at chemically labile sites of a protein and may offer a harsher alternative proteolysis method when conventional enzymes fail to cleave. Many of these chemical proteolysis methods benefit immensely from being mediated through microwave assistance and have become part of the work flow for the characterization of proteolytic enzyme-resistant proteins.

5.1 Traditional Protocols for the Chemical Digestion of Proteins

Prior to the maturation of recombinant DNA technology and protein purification technologies which allowed the production of recombinant high-grade proteomic enzymes, chemical-mediated proteolysis for protein mapping and characterization was routinely employed. These chemical proteolysis methods have changed very little over the past 50 years, and are still commonly employed for cleavage of enzyme-resistant proteins and biostructures or for cleavage at amino acid motifs that do not conform to enzymatic proteolysis sites.

Selective cleavage at aspartyl–prolyl bonds in proteins by gentle heating in the presence of acid was first reported in the 1960s.[1] It was observed that the peptide bond on the C-terminus of aspartyl groups was cleavable in dilute acid

Microwave-Assisted Proteomics
By Jennie Rebecca Lill
© Jennie Rebecca Lill 2009
Published by the Royal Society of Chemistry, www.rsc.org

solution with at least 100 times more efficiency than for other amide bonds in the same protein.[2] This selective hydrolysis could be achieved by heating the sample in either 0.03 M HCl or 0.25 M acetic acid for 5–18 h at 110 °C.[2] The aspartyl–prolyl peptide bond, in particular, was found to be exceptionally labile, and could be hydrolyzed under conditions in which other aspartyl peptide bonds were stable.[3–5] Many biomolecules possess this motif, although at a low frequency compared to other proteolytic recognition sites.

Almost a decade after aspartyl–prolyl cleavage was reported, the cleavage of proteins using cyanogen bromide (CNBr) was developed. McMenamy *et al.* demonstrated in human serum albumin and bovine α-lactalbumin that CNBr could cleave specifically at methionine residues with the cleaved methionine residue being converted to homoserine.[6] This protocol became standardized, and after preliminary optimization CNBr could hydrolyze peptide bonds C-terminal to methionine residues with greater than 90% efficiency, except where a serine or threonine residue follows methionine in the amino acid sequence. To increase the efficiency of this catalysis even further, heating of the CNBr proteolytic reaction in the presence of acidic aqueous medium was introduced, allowing in the majority of cases complete digestion at methionine and occasionally at tryptophan.[7] Several other chemical methods were also developed for proteolysis at tryptophan residues including incubation with *N*-bromosuccinimide or *N*-chlorosuccinimide; however, these reactions were occasionally prone to undesirable side reactions and formation of by-products.[8,9]

There are many chemical cleavage methods for the characterization of proteins, the most common of which are summarized in Table 5.1.[4,5,7,9–11]

After completion of the human genome project and the availability of additional genomes in database format, a major shift in the flow of proteomic research was observed. Previously protein characterization was performed by digestion of proteins followed by *de novo* sequencing using Edman degradation even up until the late 1980s where protein characterization began to focus on database searching and trypsin became the most commonly employed proteolytic tool for generating peptides for bottom-up analyses.[12] There are, however, many examples where alternative cleavage sites to trypsin, or indeed

Table 5.1   A summary of some commonly employed chemical protocols for select proteolysis and predesignated amino acids.

Proteolysis site[a]	Reagents employed
Met and Trp	CNBr in dilute acid[7]
Trp	*N*-bromosuccinimide[8] or *N*-chlorosuccinimide[9]
Asp	pH < 2; temperature > 140 °C[15]
Asp/Pro	0.03 M HCl or 0.25 M acetic acid[4,5]
Asn/Gly	2 M hydroxylamine.HCl, 2 M guanidine.HCl, 0.2 M KCO₃, pH = 9.0[10]
Cys	2-Nitro-5-thiocyanobenzoate (NTCB)[11]
General acid-labile sites	HCl, formic acid and trifluoroacetic acid[20]

[a]Met, methionine; Trp, tryptophan; Asp, aspartic acid; Pro, proline; Asn, asparagine; Gly, glycine; Cys, cysteine.

other commonly employed proteolytic enzymes, are required and where the conventional bottom-up work flow is not sufficient.

De novo sequencing, for example, is imperative for the characterization of proteins from species for which the genome is not yet deciphered,[13] or even in the analysis of human proteins, there are multiple situations where the genomic sequence does not allude to the final amino acid sequence.[14] Immunoregulatory molecules such as immunoglobulins are difficult to sequence using conventional methods that employ database searching as the protein sequence is not directly inscribed in the genome due to postprocessing events such as somatic hypermutation or gene conversion.[15] Out of the estimated 1×10^{11} antibody sequences posed by each individual, only a few thousand sequences are present in the most commonly employed databases.[16] *De novo* sequencing may also be required for the characterization of mutation sites in proteins, which again may not be contained within a curated database.

Occasionally researchers need to characterize proteins or biological samples that are impervious to enzymatic proteolytic cleavage. For example, bacterial and fungal spores have evolved to be resilient to proteolytic cleavage and therefore harsher techniques, such as chemical proteolysis, need to be employed for the generation of peptides small enough for bottom-up characterization.[17] Membrane proteins are also intrinsically resistant to tryptic characterization work flows as they suffer from solubility problems; hence chemical digestion protocols can aid in the characterization of these proteins which are typically resolubilized in non-enzymatic friendly conditions such as high concentrations of urea, salt or detergent. In addition to these examples, proteins may sometimes contain an under- or overabundance of enzymatic proteolytic cleavage sites. For example, histones are a hot topic in biomedical research at present, as histone modifications are epigenetic phenomena playing a critical role in altering chromatin structure and hence gene expression regulation in eukaryotic cells. Histones typically possess a high frequency of arginine and lysine residues; therefore tryptic digestion is not ideal for characterizing these biomolecules. In addition to this high basic amino acid frequency, these molecules also exhibit an extraordinarily high frequency of modified amino acid residues resulting in miscleavages at these sites as most proteolytic enzymes do not recognize these modified residues.[18] For biomolecules such as histones, chemical digestions would offer an attractive alternative method for characterization.

As many of the chemical proteolysis methods require heating at elevated temperatures, microwave assistance is a somewhat obvious progression in the evolution of these methods. Many groups have explored this option and some of the studies investigating microwave-assisted chemical digestion are described below.

5.2 Acid-Mediated Microwave-Assisted Digestion of Proteins

Chapter 6 summarizes how microwave-assisted acid hydrolysis (MAAH) can be employed for either N- or C-terminal sequencing, or as a tool for the complete hydrolysis of proteins into constituent amino acids for protein quantitation.[19] However, as mentioned above, acid-mediated cleavage can be performed to give proteolysis at specific amino acid residues or motifs when an enzyme is not available. In addition, chemical proteolysis is an attractive method when enzymes are not strong enough for efficient cleavage. Several acid-mediated digestion methods have recently been adopted and optimized by incorporating microwave assistance to increase yields and decrease reaction times for more efficient digestion of proteins.

5.2.1   TFA-Mediated Microwave-Assisted Proteolysis

In 2005 Zhong *et al.* demonstrated the use of MAAH for rapid protein degradation at acid-labile sites for protein identification, the main application here being the analysis of membrane proteins which are resistant to traditional tryptic digestions.[20] There are several advantages of using controlled acid hydrolysis for the proteolysis of proteins over conventional enzyme-mediated catalysis, including that the protein can be present in any solvent (*i.e.* directly in acidic solution), and does not require any special buffers for catalysis. This is particularly useful for the analysis of membrane-associated proteins since they are often solubilized in concentrated urea or harsh detergents and salts. Zhong *et al.* employed a 25% trifluoroacetic acid (TFA) aqueous solution with 20 mM dithiothreitol (DTT) to minimize oxidation of methionine, tryptophan and other amino acids with microwave irradiation for the proteolysis of protein or protein mixtures. A domestic 900 W (2450 MHz) microwave oven with a beaker of water to absorb excess heat was employed (see Chapter 2 for a discussion on microwave instrumentation and methods design). TFA microwave-mediated proteolysis was demonstrated on the hydrophobic protein bacteriorhodopsin, and resultant peptides were separated with reversed-phase HPLC, followed by analysis on a matrix-assisted laser desorption ionization (MALDI) MS/MS using a QSTAR MALDI Q*q*-TOF mass spectrometer with data interpretation using the Mascot search algorithm (Matrix Science, London, UK). The work flow for the characterization of proteins using MAAH is shown in Figure 5.1.[20]

TFA-mediated proteolysis of proteins was generally complete within 10 min, whereas in the comparative control experiment, conducted by hydrolyzing protein in 25% TFA in a conventional oven at 110 °C for 4 h, only one peak corresponding to the N-terminus was observed. During conventional heating, protein aggregation was observed which prevented further hydrolysis and therefore resulted in minimal protein cleavage. Using the microwave-mediated reaction, aggregation was not observed and efficient proteolysis progressed.

By varying the concentration of TFA (1–3 M) and irradiation time (1–10 min), the extent of proteolysis was easily controlled. Lower acid concentration (*e.g.* 0.3 M TFA) and shorter irradiation times (*e.g.* 2 min) resulted in fragments containing predominantly the N- and/or C-terminus of the protein. Higher acid concentration and lower irradiation time resulted in more internal fragment ions

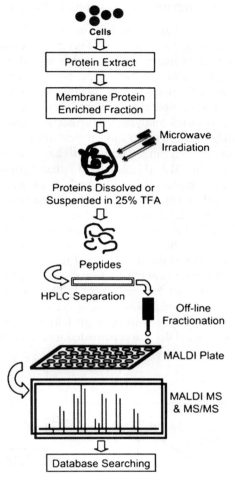

Figure 5.1 Work flow for the analysis of enriched membrane proteins by microwave-assisted acid hydrolysis for protein degradation, HPLC separation of peptides, and MALDI MS/MS with database search for protein identification.[20] (Reproduced with permission from Elsevier.)

in addition to N- and C-terminal fragment ions. Upon MAAH of the hydrophobic protein bacteriorhodopsin in the presence of 25% TFA, a labile cleavage site on both sides of glycine residues was observed suggesting this may be a particularly acid-labile residue motif. This TFA-mediated microwave-assisted method has since been employed for the structural elucidation of a cyclic peptide, kalata B2 from *Oldenlandia affinis*, by mass spectrometry. Proteolytic enzymes had previously shown insufficient cleavage of this peptide; however, TFA-mediated microwave-catalyzed proteolysis produced adequate degradation of the peptides to allow full mass spectrometric characterization.[21]

5.2.2 Asp-Specific Microwave-Assisted Proteolysis

In the introduction to this chapter, utilization of acid for aspartic acid (Asp)-specific proteolytic cleavage was discussed. This method was optimized further by Hua *et al.* whereby microwave assistance was introduced to increase catalysis of proteins at Asp-specific residues.[22] Incubation times were reduced from 8 h at room temperature to 30 s with microwave assistance in the presence of 2% formic acid, with comparable cleavage products observed by MALDI time-of-flight (TOF) mass spectral analysis. A further increase in microwave irradiation time to 6 min resulted in fewer missed cleavage products and lower mass fragment ions for peptide mapping. Zhong *et al.* had also previously examined the effect of acid type on MAAH and found that formic acid could cause modifications of some peptides during the hydrolysis process. Goodlett *et al.* also observed formylation when formic acid was employed to catalyze proteolytic reactions with modification mainly observed at the side chains or serine and threonine residues.[23] This, however, was not typically observed in the study performed by Hua *et al.*[22] The mechanism of microwave-assisted Asp-specific acid proteolysis of peptides is summarized in Figure 5.2.

Microwave-assisted formic acid cleavage was employed for the rapid identification of spore proteins from *Bacillus* spores.[15] This method was selected over conventional enzymatic approaches because spore proteins proved resistant to enzymatic proteolysis and resulted in poor protein coverage upon analysis by mass spectrometry. Chemical digestion provided increased flexibility due to the

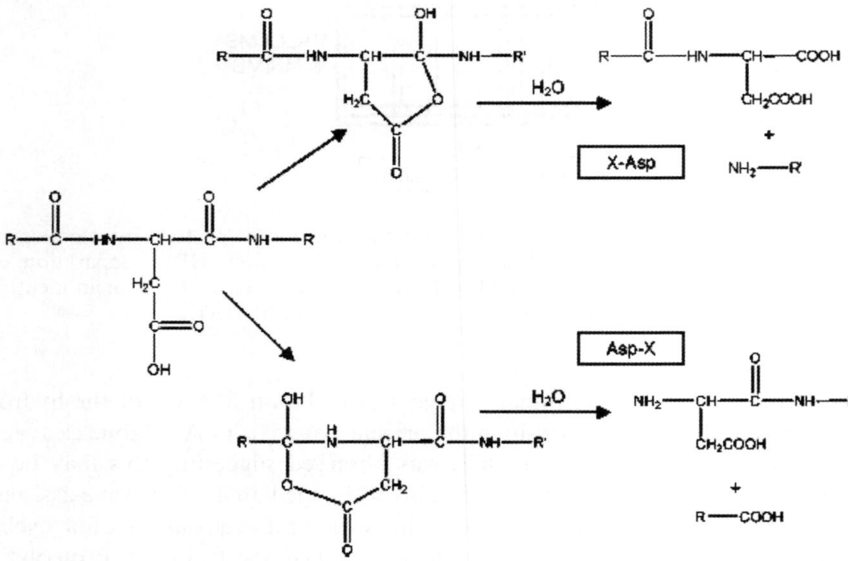

Figure 5.2   Reaction mechanisms proposed for acid-catalyzed proteolysis at pH ≤ 2 and ≥ 105 °C.[25] (Reproduced with permission from the American Chemical Society.)

harsh digestion conditions with cleavage specificity for aspartic acid residues. The *Bacillus* spores were exposed for 90 s to microwave irradiation in the presence of 6% formic acid. Samples were separated with reversed-phase HPLC on a C18 column and analyzed with a Kratos Axima-CFR Plus MALDI-TOF mass spectrometer, using post-source decay (PSD). PSD data were searched using Mascot. Signature ions from the N- and C-termini of the protein in addition to Asp-specific cleavages were observed and provided confident identification of the *Bacillus* spore proteins. It was noted during this study that, although the site of cleavage specificity was accurate, overall efficiency of digestion was relatively low resulting in the molecular weights of both the intact protein and proteolysis products being observed during MALDI-TOF analysis. This mixture of starting material and proteolysis products, however, proved beneficial as it allowed greater confidence in the characterization of *B. subtilis* spore proteins.

Swatkoski *et al.* further explored the utility of microwave-assisted acid proteolysis and its integration into proteomic work flows in a preliminary study analyzing the yeast ribosome proteome[24] and in a second study using ovalbumin and other polypeptide standards.[25] Table 5.2 summarizes the experimentally observed Asp-specific peptides from ovalbumin after acetic acid-mediated microwave-assisted hydrolysis. The majority of peptides characterized are a result of either N- or C-terminal Asp-specific cleavage, or degradation products of these peptides where N- and C-terminal "clipping" is observed.

During these studies it was demonstrated that posttranslational modifications such as phosphorylation and N-terminal acetylation were stable under microwave-assisted acid cleavage conditions, therefore increasing the universal applicability of this method to more global proteomic projects. No artifactual acetylation was observed when acetic acid was employed; however, regardless of acid type, N-terminal pyroglutamate formation occurred at a high frequency when the N-terminal amino acid residue consisted of a glutamine. For Asp-specific cleavage $pH \leq 2$ and a temperature of 140 °C in the microwave system were recommended.[23]

Hauser *et al.* also demonstrated how the non-enzymatic digestion of protein by microwave-assisted Asp cleavage is an effective proteomic tool.[26] As the Asp-specific cleavage typically leads to inherently large peptides (15–25 amino acids) that are usually relatively highly charged (greater than $+3$) when ionized by electrospray or nanospray ionization, this renders these peptides ideal for electron transfer dissociation (ETD) for tandem mass spectrometric characterization. ETD has been previously shown to give higher amino acid sequence coverage of a peptide using MS/MS than the traditional collision-induced dissociation (see Chapter 4) when performed on large more highly charged (greater than $+3$) peptides. Using ETD for the analysis of peptides generated by microwave Asp cleavage, Hauser *et al.* concluded that this allows a rapid (< 6 min) and residue-specific alternative method to enzymatic cleavage with Lys-C or Asp-N.

Sandoval *et al.* also investigated microwave-assisted acid-mediated cleavage of proteins for bottom-up protein characterization and as a tool for N- and

Table 5.2 List of experimentally observed Asp-specific peptides from ovalbumin, as assigned by Mascot.[25] (Reproduced with permission from the American Chemical Society.)

[M+H]calc	[M+H]+obs	AA Position	Amino Acid Sequence
1182.2	1182.4	350–360, 351–361	(D)AASVSEEFRA(D)
1067.2	1067.2	351–360	AASVSEEFRA
1335.4	1399.5**+Na	1–13	GSIGAASMEFCFD
1564.8	1565.3	47–59, 48–60	(D)STRTCINKVVRF(D)
1449.8	1450.2	48–59	STRTQINKVVRF
1861.9	1942.3*	336–356	AGREVVGSAEAGVDAASVSEE
2279.6	2280.5	47–66, 48–67	(D)STRTQINKVVRFDKLPGFG(D)
2164.9	2165.3	48–66	STRTQINKVVRFDKLPGFG
2597.1	2598.1	167–189, 168–190	(D)SQTAMVLVNAIVFKGLWEKAFK(D)
2482.1	2484.3	168–189	SQTAMVLVNAIVFKGLWEKAFK
2719.4	2718.9	382–385	HPFLFCIKHIATNAVLFFGRCVSP
3155.5	3139.6	139–166	Pyro-EARELINSWVESQTNGIIR NVLQPSSVD
3040.5	3024.3	139–165	Pyro-EARELINSWVESQTNGIIR NVLQPSSV
3882.6	3882.4	351–385	AASVSEEFRADHPFLFCIK HIATNAVLFFGRCVSP
4555	4556, 4636*	304–349, 305–350	(D)VFSSSANLSGISSAESLKISQAVHAAHAEINEAGREVVGSAEAGV(D)
4440	4441	305–349	VFSSSANLSGISSAESLKISQAVHAAHAEINEAGREVVGSAEAGV

5155	5198**	1–47	GSIGAASMEFCFDVFKELKV HHANENIFYCPIAIMSALAM VYLGAKD
5040	5082**	1–46	GSIGAASMEFCFDVFKELKV HHANENIFYCPIAIMSALAM VYLGAK
5384.3	5384	13–59, 14–60	(D)VFKELKVHHANENIFYCPI AIMSALAMVYLGAKDSTRTQINKVVRF(D)
5718.1	5716	304–360, 305–361	(D)VFSSSANLSGISSAESLKISQAVHAAHAEINEAGREVVGSAEAGVDAASVSEEFRA(D)
5603.1	5602	305–360	VFSSSANLSGISSAESLKISQAVHAAHAEINEAGREVVGSAEAGVDAASVSEEFRA
5848.6	5847	138–190	DQARELINSWVESQTNGIIR NVLQPSSVDSQTAMVLVNAIVFKGLWEKAFKD
6454	6454	190–246	DEDTQAMPFRVTEQESKPVQMMYQIGLFRVASMASEKMKILELPFASGTMSMLVLLP
6210	6210	192–246, 193–247	(D)TQAMPFRVTEQESKPVQMMYQIGLFRVASMASEKMKILELPFASGTMSMLVLLP(D)
6095	6095	193–246	TQAMPFRVTEQESKPVQMMYQIGLFRVASMASEKMKILELPFASGTMSMLVLLP
6701	6742**	1–60	GSIGAASMEFCFDVFKELKV HHANENIFYCPIAIMSALAM VYLGAKDSTRTQINKVVRFD
6586	6628**	1–59	GSIGAASMEFCFDVFKELKV HHANENIFYCPIAIMSALAM VYLGAKDSTRTQINKVVRFD
7416	7456**	1–67	GSIGAASMEFCFDVFKELKV HHANENIFYCPIAIMSALAM VYLGAKDSTRTQINKVVRFD KLPGFGD
7301	7341**	1–66	GSIGAASMEFCFDVFKELKV HHANENIFYCPIAIMSALAM VYLGAKDSTRTQINKVVRFD KLPGFG
8137	8135	96–166, 97–167	(D)VYSFSLASRLYAEERYPILPEYLQCVKELYRGGLEPINFQTAADQARELINSWVESQT-NGIIRNVLQPSSV(D)
8138	8217*	67–137, 68–138	(D)SIEAQCGTSVNVHSSLRDI LNQITKPNDVYSFSLASRLY AEERYPILPEYLQCVKELYR-GGLEPINFQTAA(D)

Table 5.2 (*Continued*).

[M + H]calc	[M + H] + obs	AA Position	Amino Acid Sequence
8417	8414	304–385	(D)VFSSSANLSQISSAESLKISQAVHAAHEINEAGREVVGSAEAGVDAASVSEEFRADHPFLF-CIKHIATNAVLFFGRCVSP
9033	9030	167–246, 168–247	(D)SQTAMVLVNAIVFKGLWEKAFKDEDTQAMPFRVTEQESKPVQMMYQIGLFRVASMASE-KMKILELPFASGTMSMLVLLP(D)
9917	9915	168–246	SQTAMVLVNAIVFKGLWEKAFKDEDTQAMPFRVTEQESKPVQMMYQIGLFRVASMASE-KMKILELPFASGTMSMLVLLP

Table 5.3 A comparison of the different acid-mediated proteolysis methods described in the literature. (Adapted from Hauser *et al.*[26])

Conditions and microwave exposure	Cleavage pattern
6 M HCl (30 s–2 min)	Predominantly N- and C-termini[19]
0.1 M HCl or 0.3 M TFA (1–2 min)	Predominantly N- and C-termini[20]
25% TFA (10 min)	Internal cleavage at predominantly acid-labile sites
1 M TFA (2 min)	Predominantly N- and C-termini
2% formic acid (10 min)	Cleavage specifically at aspartic acid[22,24]
Formic acid or acetic acid	Cleavage at N-termini, but often see modification (formylation/acetylation)[17,20,24]

C-terminal sequencing.[27] The findings of these investigations and a summary of the cleavage sites reported from the literature are summarized in Table 5.3.

In summary, MAAH was proven to be a useful tool for rapid protein identification. The TFA and formic acid methods described above produced peptides of analogous peak sizes to tryptic and Asp-N digestions respectively and allowed cleavage of highly enzymatic-resistant proteins. Refer to protocols VI and VII in Chapter 10 for practical advice on performing these digestions.

5.3 Other Microwave-Assisted Chemical Proteolysis Reactions

To date there have been no reports on applying microwave assistance for CNBr, asparagine/glycine or other chemical cleavage protocols. However, in our laboratory microwave assistance has been investigated for increased catalysis of several chemical proteolytic reactions, following the simple rule that if the reaction is typically carried out at elevated temperatures in the oven, it cannot be detrimental to perform these reactions in the microwave (unless the substrate protein is particularly heat labile). One suggestion as to why these other reactions have not been publicized in the literature is that they are often hard to quantify, as on test proteins the reactions can occur in a very short space of time, and in most cases the non-microwave-mediated reaction is already fast. For any proteins that appear to be resilient to oven-mediated digestions, or indeed for which no proteolytic enzyme is available for digestion at "convenient sites", microwave-mediated chemical digestions are ideal. When performing these chemical digestions, the researcher should pay particular attention to safety, as many of the reagents listed in Table 5.1 (in particular CNBr) are potent chemicals with carcinogenic and toxic properties. Care must be taken not to inhale or become exposed to aerosolized chemicals during the rapid heating process produced by microwave-mediated reactions, and it is prudent to place the microwave unit in a well-ventilated fume-hood. In addition, many of the reagents summarized in Table 5.1 are also prone to decomposing in solution (again, in particular CNBr) and therefore fresh reagents should be employed for each new analysis.

In summary, microwave-assisted chemical proteolysis has been demonstrated as a useful tool for protein characterization and global proteomic research. The majority of chemical cleavages result in longer peptides than average tryptic peptides and these longer peptides can be better resolved chromatographically[21,22] and also may carry higher charge states leading to more efficient and informative fragmentation patterns than significantly shorter peptides. In addition, chemical-mediated microwave-assisted digestions offer an attractive solution for proteins that have proved resilient to traditional proteolytic digestion techniques.

References

1. J. Shultz, *Methods Enzymol.*, 1967, **11**, 255.
2. A. Light, *Methods Enzymol.*, 1967, **11**, 417.
3. D. Piszkiewicz, M. Landon and E. L. Smith, *Biochem. Biophys. Res. Commun.*, 1970, **40**, 1173.
4. M. Landon, *Methods Enzymol.*, 1977, **47**, 145.
5. F. Marcus, *Int. J. Peptide Protein Res.*, 1985, **25**, 542.
6. R. H. McMenamy, H. M. Dintzis and F. Watson, *J. Biol. Chem.*, 1971, **246**, 4744.
7. R. Kaiser and L. Metzka, *Anal. Biochem.*, 1999, **266**, 1.
8. L. K. Ramachandran and B. Witkop, *Methods Enzymol.*, 1976, **11**, 283.
9. M. A. Lishwe and M. T. Sung, *Anal. Biochem.*, 1977, **127**, 453.
10. P. Bornstein and G. Balian, *Methods Enzymol.*, 1977, **47**, 132.
11. R. G. Stark, *Methods Enzymol.*, 1977, **47**, 129.
12. J. V. Olsen, S. E. Ong and M. Mann, *Mol. Cell Proteomics*, 2004, **3**, 608.
13. N. Bandeira, K. R. Clauser and P. A. Pevzner, *Mol. Cell Proteomics*, 2007, **6**, 1123.
14. V. Pham, W. J. Henzel, D. Arnott, S. Hymowitz, W. N. Sandoval, B. T. Truong, H. Lowman and J. R. Lill, *Anal. Biochem.*, 2006, **352**, 77.
15. J. M. Di Noia and M. S. Neuberger, *Annu. Rev. Biochem.*, 2007, **76**, 1.
16. E. A. Padlan and E. A. Kabat, *Methods Enzymol.*, 1991, **203**, 3.
17. S. Swatkoski, S. C. Russell, N. Edwards and C. Fenselau, *Anal. Chem.*, 2006, **78**, 181.
18. B. Ueberheide and S. Mollah, *Int. J. Mass Spectrom.*, 2007, **259**, 46.
19. H. Zhong, Y. Zhang, Z. Wen and L. Li, *Nat. Biotechnol.*, 2004, **22**, 1291.
20. H. Zhong, S. L. Marcus and L. Li, *J. Am. Soc. Mass Spectrom.*, 2005, **16**, 471.
21. S. S. Nair, J. Romanuka, M. Billeter, L. Skjeldal, M. R. Emmett, C. L. Nilsoon and A. G. Marshall, *Biochim. Biophys. Acta*, 2006, **1764**, 1568.
22. L. Hua, T. Y. Low and S. K. Sze, *Proteomics*, 2005, **6**, 586.
23. D. R. Goodlett, F. B. Armstrong, R. J. Creech and R. B. Van Breemen, *Anal. Biochem.*, 1990, **186**, 116.

24. S. Swatkoski, P. Gutierrez, J. Ginter, A. Petrov, J. D. Dinman, N. Edwards and C. Fenselau, *J. Proteome Res.*, 2007, **6**, 4525.
25. S. Swatkoski, P. Gutierrez, J. Ginter, A. Petrov, J. D. Dinman, N. Edwards and C. Fenselau, *J. Proteome Res.*, 2008, **7**, 579.
26. N. J. Hauser, H. Han, S. A. McLuckey and F. Basile, *J. Proteome Res.*, 2008, **7**, 1867.
27. W. N. Sandoval, V. Pham, E. S. Ingle, P. S. Liu and J. R. Lill, *Comb. Chem. High Throughput Screening*, 2007, **10**, 751.

CHAPTER 6

Microwave-Assisted Acid Hydrolysis

Abstract

Microwave-assisted acid hydrolysis of proteins can be employed for a number of analytical procedures including the hydrolysis of single recombinant proteins into constituent amino acids for quantitation, as an alternative tool for N- and C-terminal sequencing and for protein digestion for bottom-up protein characterization as previously discussed in Chapter 5. This chapter describes the evolution of these techniques and their applications in the characterization and quantitation of proteins.

6.1 Protein Quantitation by Amino Acid Analysis

Protein quantitation is an extremely important tool for many drug discovery and analytical protocols including *in vivo* drug dosing experiments, protein interaction and X-ray crystallography studies, and also for various *in vitro* investigations. There are several methods that allow accurate quantitation of protein concentration including traditional approaches such as the bicinchoninic acid (BCA) protein assay, the Bradford protein assay, the Lowry protein assay and the non-absorbant/colorimetric assay, amino acid analysis assay (AAA). The first step in AAA involves hydrolyzing the protein to its constituent amino acids typically by breaking the amide bonds in the presence of 6 N HCl at 110 °C for 24 h, a method first described by Hirs *et al.* over 50 years ago.[1]

Resulting hydrolyzates are analyzed and quantitated against standard amino acids that have been injected at predefined amounts. AAA methods can be categorized as either pre- or post-derivatization. This refers to the labeling of

Microwave-Assisted Proteomics
By Jennie Rebecca Lill
© Jennie Rebecca Lill 2009
Published by the Royal Society of Chemistry, www.rsc.org

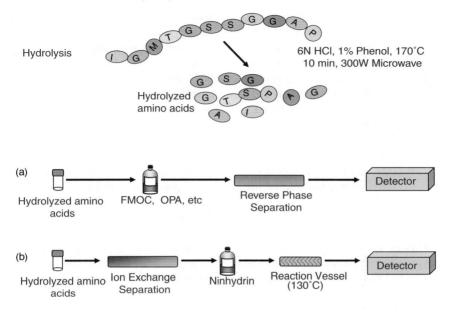

Figure 6.1 Work flow for (a) pre-column and (b) post-column derivatization amino acid analysis for protein quantitation.

the amino acids either before or after a chromatographic separation step. Although several new technologies have recently become available for performing AAA such as the incorporation of ITRAQ multiplexing labels[2] and the incorporation of UPLC separations,[3] traditional methods still rely on the use of either (a) *o*-phthalaldehyde (OPA) and 9-fluoromethylchloroformate (FMOC) or a number of other commercially available reagents for pre-column derivatization, or (b) ninhydrin for post-column derivatization. Figure 6.1 shows a schematic of these two commonly employed AAA work flows.

Traditionally proteins are hydrolyzed in the vapor phase with 6 N HCl at 110 °C for 24 h,[1] although this can be performed at elevated temperatures for just 1 h at the expense of compromising the hydrolysis vessel.[4] As analysis times for hydrolyzates typically range between 12 and 90 min, this lengthy hydrolysis time is often the rate-limiting step in AAA. Microwave-assisted acid hydrolysis (MAAH) has therefore been investigated and adopted by many groups to allow the rapid hydrolysis of samples and therefore significantly reducing analysis time.

6.2 Microwave-Assisted Acid Hydrolysis for Protein Quantitation

Protein hydrolysis for AAA was one of the preliminary protein chemistry methods shown to benefit from microwave assistance and there are several comprehensive reports in the literature showing the benefits of microwave-assisted protein hydrolysis.[5–8] The first study investigating the potential of MAAH can be attributed to Gilman and Woodward who studied microwave irradiation for the vapor-phase hydrolysis of proteins.[5] In this initial study, methionyl human growth hormone (m-HGH Protropin®) was hydrolyzed in a model MDS-81D microwave digestion system (CEM Corp.) in a Teflon PFA digestion vessel (CEM Corp.) filled with 10 mL of 6 N HCl. Microwave parameters were set at 650 W and 965 kPa (140 psig). Stable temperature (178–180 °C) was achieved by setting the microwave pressure controller. Accuracy and precision of replicate 8–12 min microwave-assisted vapor-phase hydrolyzates were equivalent to typical 24 h, 110 °C heating block hydrolyzates. One problem with conventional vapor-phase protein hydrolysis is that the vessel can be prone to leakage. Leakage can lead to lower recovery of hydrolysis products (*i.e.* amino acids) and is often not detected until the final analysis of resultant hydrolyzate products. With microwave-mediated hydrolysis in the commercially available microwave apparatus mentioned above, leaks could easily be detected prior to the final analysis step, due to a drop in the monitored pressure, therefore again reducing troubleshooting time if a leak should occur.[5]

Concurrent with the investigations by Gilman and Woodward, Chiou and Wang proposed a microwave-assisted protein hydrolysis method that could be performed in just 4–12 min.[6] The risk of explosion from performing acid hydrolyses at high pressure and temperature was demonstrated, and therefore Teflon/Pyrex custom-made vials were introduced as safer vessels in which to carry out MAAH.

Several other researchers[7,8] further optimized the protocol for MAAH for the complete hydrolysis of proteins into constituent amino acids with the aim to have complete hydrolysis in a uniform manner on batches of samples. Davidson summarized the various hydrolysis protocols for amino acid analysis and proposed a final protocol hydrolysis in a model MDS-81D microwave digestion system (CEM Corp.). The power was set at 650 W for 8 min and the protocol allowed more uniform hydrolysis conditions than previously cited procedures.[8]

Sandoval *et al.* published data on a comparison between a non-microwave-based method described by Higbee *et al.*,[9] the standard protein hydrolysis method in their laboratory, and a commercially available protein hydrolysis method marketed by CEM.[10] Samples were hydrolyzed in an evacuated chamber with 6 N HCl with 0.1% phenol at 110 °C for 24 h or 150 °C for 1 h in a conventional convection oven, or at 175 °C for 10 min at 300 W in the Discover Microwave (CEM, Newark, CA). The hydrolyzed samples were then analyzed using a Hitachi L-8800 amino acid analyzer (Hitachi High Technologies America, San Jose, CA).

The results obtained using the microwave-assisted technique were comparable to much lengthier traditional hydrolysis protocols. Figure 6.2 shows the recoveries of amino acids from hydrolyzed β-casein with theoretical amino acid recoveries. Overall, comparable amino acid recoveries were observed for proteins hydrolyzed by microwave irradiation at 175 °C for 10 min compared to proteins that were hydrolyzed using two common protocols, *i.e.* non-microwave-mediated

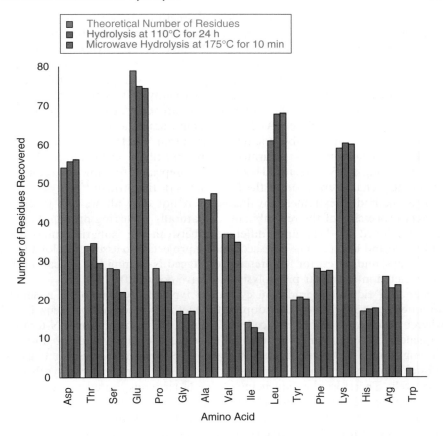

Figure 6.2 Comparison of amino acid recoveries after hydrolysis using microwave-mediated hydrolysis and two conventional oven-based hydrolysis methods.[4] (Reproduced with permission from *Combinatorial Chemistry & High Throughput Screening*.)

incubation at 110 °C for 24 h or at 150 °C for 1 h.[4] This protocol is summarized in protocol V in Chapter 10.

This microwave-assisted technique has now been validated on the quantitation of thousands of recombinant proteins and proves an invaluable tool for the rapid quantitation of proteins in the biotechnology industry.

6.3 Traditional Methods for N- and C-Terminal Sequencing

In the biotechnology industry it is important to verify that correct transcriptional and intracellular processing of a recombinant protein has occurred. N-terminal sequencing can be performed using several chemistries including the

Sanger method[11] and the dansyl chloride[12] method; however, the most conventional protocol currently employs chemistries first described by Edman using derivatization of the free N-terminal amine with phenylisothiocyanate.[13] The first step of Edman degradation is termed coupling; here, the free ε-amino group of the N-terminus of the protein is derivatized with phenylisothiocyanate under basic conditions to form a phenylthiocarbamyl (PTC) product. This PTC amino acid is cleaved from the protein by trifluoroacetic acid (TFA) to form an unstable anilinothiazolinone (AZT) intermediate amino acid which is converted immediately to a more stable phenylthiohydantoin (PTH) amino acid. It is this PTH amino acid that is separated by reverse-phase chromatography and detected using a UV detector. This process is repeated to allow the sequential amino acid characterization of the N-terminus of the protein.[13]

This method is extremely invaluable as it not only allows one to ensure correct processing of the recombinant or naturally occurring protein, but it is also employed to distinguish differences between the isobaric amino acids isoleucine and leucine in *de novo* sequencing projects[14] to assess the clonality of antibodies, and to test for the presence of ragged N-termini (an indication that unpredicted intracellular proteolysis may have occurred).

Standard Edman degradation cycles are typically 30 min to 1 h in length, although several groups have adopted the methods described by Henzel *et al.* whereby each cycle can be reduced to 12 min.[15] For a conventional N-terminal sequencing sample a minimum of 12 cycles are required, and for antibody cloning studies between 25 and 30 cycles are typically necessary to gather enough information for a polymerase chain reaction primer to be synthesized effectively. Therefore the sequencing of one N-terminus can take several hours to complete. In addition to the amount of time it takes to sequence one N-terminus, a further limitation of N-terminal sequencing using Edman chemistries is that it requires a free ε-amino group at the N-terminus. It has been estimated, however, that over 80% of naturally occurring mammalian proteins are N-terminally modified, rendering the N-terminus inaccessible to coupling with phenylisothiocyanate and therefore prohibiting Edman sequencing.[16]

Due to the limitations of Edman chemistries, a variety of biochemical protocols followed by mass spectrometric analysis[17,18] and top-down mass spectrometric proteomic approaches[19] have recently been described as alternative methods for N-terminal characterization. Although biochemical capture methodologies offer an alternative approach to Edman chemistries for N-terminal identification, these can be relatively lengthy protocols prone to sample loss due to the multiple biochemistries involved. The top-down approaches such as electron capture dissociation (ECD) and the methods described by Miksesh *et al.* for electron transfer dissociation (ETD), with or without incorporation of proton transfer reduction (PTR), are still in their infancy and at present are only truly accessible to uniform protein species at high levels of concentration (picomoles) of relatively low molecular weight ($<30\,kDa$).[19]

In addition to N-terminal sequencing, chemistries have also been described for C-terminal sequencing in a manner analogous to Edman degradation.[20] For these chemistries, large amounts (nanomoles) of protein are required due to the inefficient coupling, cleavage and conversion events, and consequently many researchers employ alternative C-terminal sequencing protocols such as bottom-up protein characterization, employment of controlled carboxypeptidase reactions or affinity capture methods such as anhydrotrypsin capture.[21]

6.4 N- and C-terminal Sequencing using Microwave-Assisted Acid Hydrolysis

In 2004 Zhong *et al.* described the use of MAAH as an alternative method for the N- and C-terminal characterization of proteins.[22] MAAH involves exposure of proteins to high concentrations of acid under microwave irradiation, resulting in the denaturation and cleavage of the proteins at acid-labile sites. The initial protocol described the exposure of proteins in 6 N HCl to microwave irradiation for small increments of time (30 s to 2 min) with resultant hydrolysis products analyzed using matrix-assisted laser desorption ionization (MALDI) time-of-flight (TOF) analysis. It was discovered that after performing MAAH on intact proteins for short exposure times, ions corresponding to incremental ladders of amino acid chains from the N- and C-termini of the proteins were observed. Figure 6.3 is a reproduction from the work of Zhong *et al.* showing a schematic of the mass analysis of polypeptide ladders generated from MAAH using 6 N HCl. Ions corresponding to truncations at each amino acid along the termini are observed.[22]

In Chapter 5 an MAAH protocol employing TFA for the generation of peptides by cleavage at acid-labile bonds for bottom-up mass spectrometric analysis was described.[23] During this investigation into the cleavage of proteins using TFA for bottom-up protein characterization, it was deduced that if proteins were exposed to 1.5 M TFA for a brief time period (1–2 min) then only the N- and C-terminal peaks were observed. Sandoval *et al.*[4] therefore investigated both the original HCl-mediated MAAH protocol and the TFA hydrolysis method as potential alternatives to Edman degradation for the high-throughput N-terminal sequencing of proteins. The same trends as described by Zhong *et al.* were observed and after a slight modification to the protocol, a technique employing 1 M TFA with exposure for 2 min at 100 °C at 50 W for N- and C-terminal sequence analysis of recombinant proteins was adopted. Figure 6.4 shows an example of this work whereby murine serum albumin (MSA) was subjected to MAAH using the protocol described above, with resultant peptides analyzed in full MS using MALDI-TOF with a DE-STR Voyager (PE-Sciex, Toronto, Canada).

Peaks corresponding to N-terminal truncations were observed as well as a C-terminal peptide ion. During this investigation into MAAH as an alternative method for N-terminal sequencing it was observed that one terminus from the protein was often preferentially observed in the MALDI-TOF data over the other; however, in some cases both termini were seen in equal quantities. In the case of myoglobin preferential cleavage at the N-terminus was commonly observed (Figure 6.5), whereas for lysozyme (Figure 6.6) C-terminal peaks were observed.

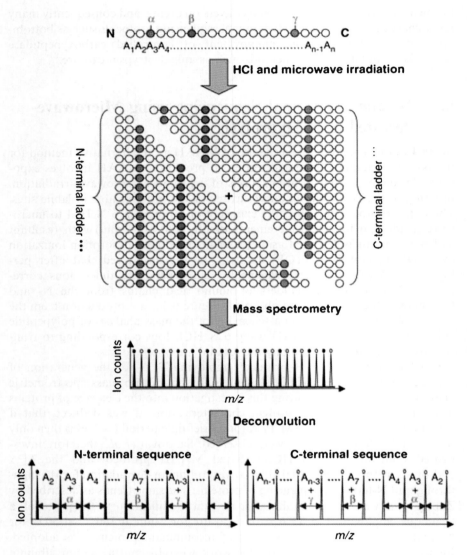

Figure 6.3   Schematic of the MAAH method for sequencing the N- and C-termini of
proteins. The mass spectrum produced consists of peaks exclusively from
the N- and C-terminal polypeptides of the protein. Amino acid sequences
and any modifications are read from the mass differences of adjacent
peaks within the same series of the polypeptide ladder.[22] (Reproduced
with kind permission from Dr Liang Li from *Nature Biotechnology*.)

It was hypothesized that the high abundance of N- and C-terminal truncations
was primarily due to acid lability of the more highly exposed termini and protein
conformation, *i.e.* the more exposed the terminus in its tertiary structure, the
more likely that it will be preferentially cleaved during MAAH.[4] Figures 6.7a

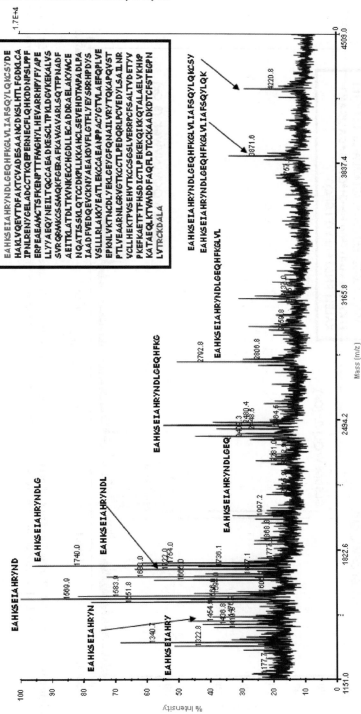

Figure 6.4 MALDI-TOF mass spectrum of peptides generated from mouse serum albumin after performing MAAH in 1 M TFA at 50 W for 2 min in a CEM Discover microwave system.[4] (Reproduced with permission from *Combinatorial Chemistry & High Throughput Screening*.)

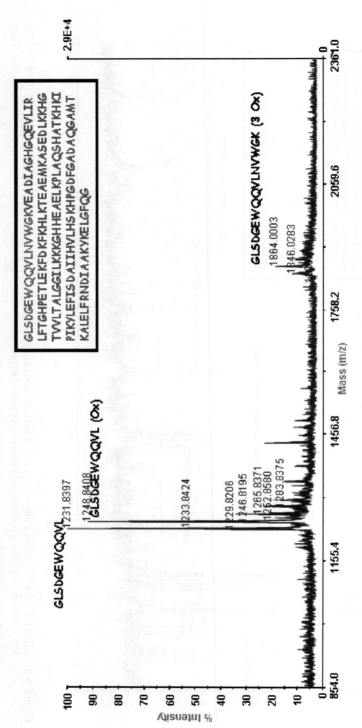

Figure 6.5 MALDI-TOF mass spectrum of peptides generated from horse myoglobin after performing MAAH in 1 M TFA at 50 W for 2 min in a CEM Discover microwave system.[4] (Reproduced with permission from *Combinatorial Chemistry & High Throughput Screening*.)

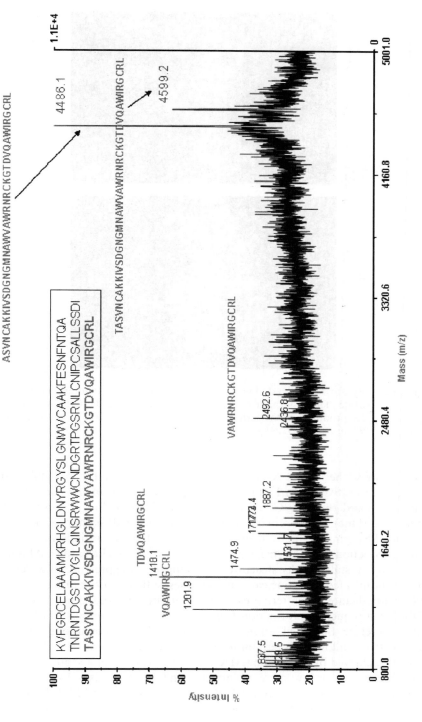

Figure 6.6 MALDI-TOF mass spectrum of peptides generated from lysozyme after performing MAAH in 1 M TFA at 50 W for 2 min in a CEM Discover microwave system.[4] (Reproduced with permission from *Combinatorial Chemistry & High Throughput Screening*.)

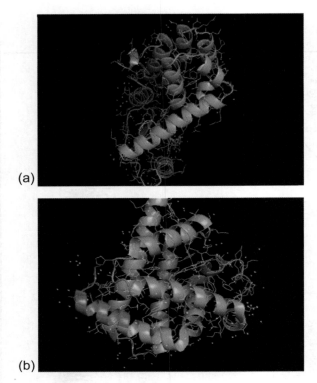

Figure 6.7 PyMol X-ray crystallography structures for (a) myoglobin and (b) lyso-zyme. The N-termini are represented in blue and the C-termini in red.[4] (Reproduced with permission from *Combinatorial Chemistry & High Throughput Screening*.)

and 6.7b show the PyMol X-ray crystallography structure for myoglobin and lysozyme, respectively.[4] In Figure 6.7a the N-terminus of myoglobin (in blue) is more exposed than the C-terminus (red). In Figure 6.7b for lysozyme, the N-terminus (blue) is buried whereas the C-terminus (red) is more exposed. In addition, lysozyme contains disulfide bonds at the N-terminus, and hence without reduction the molecule is more resistant to MAAH cleavage. Although these X-ray crystallography structures are mapped in the presence of water and not in concentrated acid, one can still hypothesize that conformation as well as acid lability dictate which terminus is preferentially observed after MAAH.[4]

Obtaining sequence information from more than one terminus can be bene-ficial. Therefore Sandoval *et al.* investigated whether conditions could be manipulated to allow both termini to be visualized in one analysis.[4] This could indeed be achieved by performing MAAH in the presence of a small percentage of organic solvent or low salt concentration and, in some cases, by performing reduction and alkylation. Figure 6.8 shows myoglobin after MAAH has been performed. In Figure 6.8a, only ions corresponding to the N-terminus are

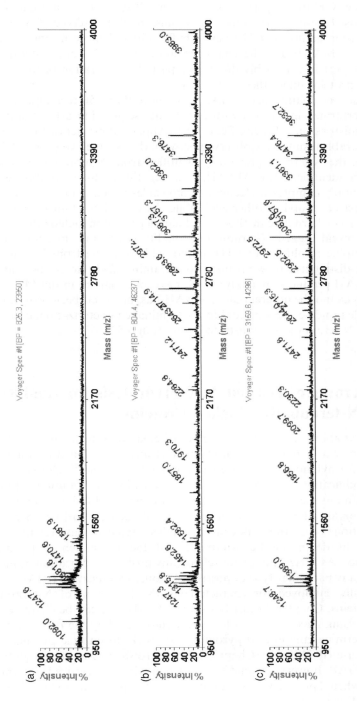

Figure 6.8 MALDI-TOF mass spectra of peptides generated from myoglobin after performing MAAH in (a) 1 M TFA at 50 W for 2 min in a CEM Discover microwave system (b) 1 M TFA with addition of 20% acetonitrile and (c) 1 M TFA with addition of 250 mM Tris at pH = 8.[4] (Reproduced with permission from *Combinatorial Chemistry & High Throughput Screening*.)

observed; however, after exposure to organic or salt solutions, as in Figures 6.8b and 6.8c, ions corresponding to the C-terminus are also observed. This manipulation by using additives in the acidic buffer allowed more flexibility in the application of MAAH for the characterization of the N- and C-termini of recombinant proteins, probably due to the partial denaturation of the tertiary structure of proteins under these conditions.

An extensive evaluation using MAAH on more than 150 recombinant proteins was performed in parallel with N-terminal sequencing using traditional Edman chemistries on a Procise Sequencer (Applied Biosystems, Foster City, CA). Comparable results were observed between the two methods for $>75\%$ of proteins, *i.e.* the N-termini previously identified from N-terminal sequencing could also be identified using MAAH; however, this excluded antibodies and other structurally complex molecules, or molecules which contained an abundance of aspartyl–prolyl or other acid-labile cleavage sites.[4] Other limitations of MAAH as an alternative method to Edman degradation include that it does not always reveal ragged N-termini or clonality issues and, in addition, the chromatographic resolution of PTH amino acids resulting from Edman chemistries can distinguish between the isobaric amino acids isoleucine and leucine; as MAAH products are analyzed by mass spectrometry in full MS mode, these species cannot be differentiated. MAAH has, however, been proven to be an invaluable rapid verification tool for N-terminal characterization, especially for blocked N-termini as demonstrated in Section 6.5.[4]

6.5 Microwave-Assisted Acid Hydrolysis for Analysis of N-terminally Blocked Proteins

As mentioned above, a significant number of eukaryotic proteins have N-termini blocked with a number of modifications including, but not limited to, acetylation, formylation, methylation and pyroglutamic acid.[16] This type of N-terminal modification renders the protein inaccessible to traditional Edman chemistries and in such cases either a deblocking protocol followed by Edman chemistries or an alternative analytical protocol such as bottom-up protein characterization is performed. Bottom-up analyses typically have relatively low success rates at identifying the N-terminus if it does not match the pre-annotated sequence. As MAAH does not rely on any protein modification or indeed bias against any particular type of posttranslational modification (PTM), it can be successfully employed for characterizing the N-termini of N-terminally blocked proteins. Chapter 7 describes several PTMs that can be characterized or removed using microwave-assisted techniques including the removal of the cyclized N-terminal amino acid pyroglutamate with the enzyme pyroglutamyl aminopeptidase (PGAP) (see Chapter 7 for microwave-assisted pyroglutamyl removal). PGAP, however, is ineffective at removing the cyclized N-terminal amino acid when a proline residue follows the initial pyroglutamyl residue, due to conformational inaccessibility.

Although other methods are typically employed to gain sequence information including biochemical capture methods or bottom-up protein characterization protocols, these methods require prior knowledge of the protein sequence so that the correct proteolytic enzyme can be employed. In many cases, the N-terminus does not contain a convenient enzymatic cleavage site making these methods cumbersome. MAAH was therefore demonstrated on a number of N-terminally blocked proteins, for example hepatocyte growth factor (HGF), the α-chain of which was N-terminally blocked with a pyroglutamate residue that, despite the absence of a proline residue at position 2, was resistant to PGAP digestion using conventional methods.[4]

Figure 6.9 shows a MALDI-TOF mass spectrum of full length HGF after MAAH.[4] After 2 min MAAH incubation at 100 °C in a microwave system followed by MALDI mass spectral analysis, ions corresponding to the

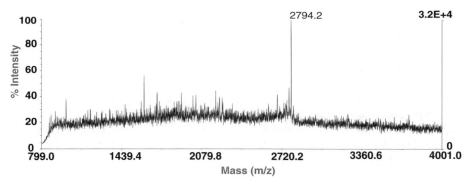

QRKRRNTIHEFKKSAKTTLIKIDPALKIKTKKVNTADQCANRCTRNK
GLPFTCKAFVFDKARKQCLWFPFNSMSSGVKKEFGHEFDLYENKDYIR
NCIIGKGRSYKGTVSITKSGIKCQPWSSMIPHEHSFLPSSYRGKDLQENY
CRNPRGEEGGPWCFTSNPEVRYEVCDIPQCSEVECMTCNGESYRGLMD
HTESGKICQRWDHQTPHRHKFLPERYPDKGFDDNYCRNPDGQPRPWC
YTLDPHTRWEYCAIKTCADNTMNDTDVPLETTECIQGQGEGYRGTVNT
IWNGIPCQRWDSQYPHEHDMTPENFKCKDLRENYCRNPDGSESPWCFT
TDPNIRVGYCSQIPNCDMSHGQDCYRGNGKNYMGNLSQTRSGLTCSM
WDKNMEDLHRHIFWEPDASKLNENYCRNPDDDAHGPWCYTGNPLIP
WDYCPISRCEGDTTPTIVNLDHPVISCAKTKQLRVVNGIPTRTNIGWMV
SLRYRNKHICGGSLIKESWVLTARQCFPSRDLKDYEAWLGIHDVHGRG
DEKCKQVLNVSQLVYGPEGSDLVLMKLARPAVLDDFVSTIDLPNYGCT
IPEKTSCSVYGWGYTGLINYDGLLRVAHLYIMGNEKCSQHHRGKVTLN
ESEICAGAEKIGSGPCEGDYGGPLVCEQHKMRMVLGVIVPGRGCAIPNR
PGIFVRVAYYAKWIHKIILTYKVPQS

Figure 6.9 MALDI-TOF spectrum after MAAH of a blocked protein, hepatocyte growth factor (HGF). The α-chain is blocked with a pyroglutamate residue. PGAP enzyme is ineffective on this protein and therefore Edman sequencing is not possible. MAAH delivered sequence information from the N-terminus of the α-chain.[4] (Reproduced with permission from *Combinatorial Chemistry & High Throughput Screening*.)

N-terminal sequence of HGF α-chain were observed, allowing fast and accurate sequence verification. MAAH was therefore proven as a useful tool for the characterization of blocked N-termini allowing higher throughput analysis than traditional deblocking or "bottom-up" methods alone.

References

1. C. H. W. Hirs, W. H. Stein and S. Moore, *J. Biol. Chem.*, 1954, **211**, 941.
2. E.S. Ingle, J.R. Lill, C.J. Bramwell, S. Daniels and S. Nimkar, Proceedings of the 54th ASMS Conference on Mass Spectrometry and Allied Topics, Seattle, WA, 2006.
3. K. Yu, P. Alden and R. Plumb, Proceedings of the 54th ASMS Conference on Mass Spectrometry and Allied Topics, Seattle, WA, 2006.
4. W. N. Sandoval, V. Pham, E. S. Ingle, P. S. Liu and J. R. Lill, *Comb. Chem. High Throughput Screening*, 2007, **10**, 751.
5. L. B. Gilman and C. Woodward, in *Current Research in Protein Chemistry: Techniques, Structure, and Function*, ed. J. J. Villafranca, Academic Press, San Diego, CA, 1990, p. 23.
6. S.-H. Chiou and K.-T. Wang, in *Current Research in Protein Chemistry: Techniques, Structure, and Function*, ed. J. J. Villafranca, Academic Press, San Diego, CA, 1990, p. 3.
7. S.-T. Chen, S.-H. Chiou and K.-T. Wang, *J. Chin. Chem. Soc.*, 1991, **38**, 85.
8. I. Davidson, in *Methods in Molecular Biology*, ed. B. J. Smith, Humana Press, Totowa, NJ, 1996, p. 119.
9. A. Higbee, S. Wong and W. J. Henzel, *Anal Biochem.*, 2003, **318**, 155.
10. J. R. Lill, E. S. Ingle, P. S. Liu, V. Pham and W. N. Sandoval, *Mass Spectrom. Rev.*, 2007, **26**, 657.
11. F. Sanger and H. Tuppy, *Biochem. J.*, 1961, **49**, 463.
12. W. R. Gray, *Methods Enzymol.*, 1972, **25**, 121.
13. P. Edman, *Eur. J. Biochem.*, 1967, **1**, 80.
14. V. Pham, W. J. Henzel, D. Arnott, S. Hymowitz, W. N. Sandoval, B. T. Truong, H. Lowman and J. R. Lill, *Anal. Biochem.*, 2006, **352**, 77.
15. W. J. Henzel, J. Tropea and D. Dupont, *Anal. Biochem.*, 1999, **267**, 148.
16. J. L. Brown and W. K. Roberts, *J. Biol. Chem.*, 1976, **251**, 1009.
17. L. McDonald, D. H. Robertson, J. L. Hurst and R. J. Beynon, *Nat. Methods*, 2005, **12**, 955.
18. S. M. Brittain, S. B. Ficcaro, A. Brock and E. C. Peters, *Nat. Biotechnol.*, 2005, **23**, 463.
19. L. M. Miksesh, B. Ueberheide, A. Chi, J. J. Coon, J. E. Syka, J. Shabanowitz and D. F. Hunt, *Biochim. Biophys. Acta*, 2006, **1764**, 1811.
20. A. S. Inglis, *Anal. Biochem.*, 1991, **195**, 183.
21. S. Sechi and B. T. Chait, *Anal. Chem.*, 2000, **72**, 3374.
22. H. Zhong, Y. Zhang, Z. Wen and L. Li, *Nat. Biotechnol.*, 2004, **22**, 1291.
23. H. Zhong, S. L. Marcus and L. Li, *J. Am. Soc. Mass Spectrom.*, 2005, **16**, 471.

CHAPTER 7

Microwave-Assisted Discovery and Characterization of Posttranslational Modifications

Abstract

The huge diversity of the proteome is attributed to several posttranslational events, in particular the presence of more than 200 covalently attached post-translational modifications (PTMs). These modifications play a critical role in controlling interactions at both a molecular and cellular level, and therefore it is important to be able to characterize these modifications in order to gain insight into the mechanisms involved within intracellular and extracellular pathways. A variety of tools are available for the analysis or characterization of PTMs; however, many of these protocols can benefit in terms of decreased reaction times or increased biochemical efficiency by performing the reactions with microwave assistance. This chapter summarizes some of the key findings from the literature for the microwave-assisted characterization of PTMs.

7.1 Posttranslational Modifications

A posttranslational modification (PTM) is a chemical modification of a protein occurring after translation, although some co-translational modifications exist which are often filed into the same category as PTMs.[1] There are a wide range of modifications that can take place, acting either on individual residues or on multiple sites within a given protein. There are, in total, over 200 different PTMs catalogued in eukaryotic and prokaryotic species[2] and this number is growing due to enhanced bioanalytical approaches and bioinformatic tools.[3,4] PTMs define the functional and structural plasticity of proteins in all species. The huge diversity of the proteome is attributed to the plethora of

Microwave-Assisted Proteomics
By Jennie Rebecca Lill
© Jennie Rebecca Lill 2009
Published by the Royal Society of Chemistry, www.rsc.org

modifications that can occur, giving each protein the possibility to take on multiple conformations and therefore cellular and biological roles. During recent years, protein PTMs have attracted attention in the biological and biomedical research communities and it is now clear that most proteins in species from archaea to humans carry site-specific covalent modifications of one type or another.

For a complete list of commonly used PTMs, the reader is referred to the following website: http://www.unimod.org.

Characterization of PTMs often proves to be a daunting task as an individual protein may contain multiple sites of modification, or indeed a heterogeneous complement of PTMs. Figure 7.1 shows a schematic of the localization and role of several of the more common PTMs observed in eukaryotic organisms.[2]

Different PTMs possess highly variable physicochemical properties. Common challenges with characterizing PTMs include the issue of stability under analytical conditions, the level of abundance, which is often at sub-stoichiometric amounts, and also complexities related to heterogeneity. Multiple methods commonly exist for the complete characterization of even just a single PTM: affinity capture protocols,[5,6] diagonal chromatography methods,[7] mass spectrometric work flows such as precursor ion scanning and neutral loss scan functions[8,9] in addition to suites of bioinformatic tools[3–4,10] which all aid in the characterization of PTMs.

This chapter focuses on several methods recently described in the literature for the characterization of PTMs using microwave-assisted methods. Some of these are enhancements to traditional biochemical protocols and others are new concepts in PTM discovery and characterization.

7.2 Microwave-Assisted Characterization of Glycopeptides and Glycoproteins

The term glycan refers to a monosaccharide or polysaccharide, and when this carbohydrate is conjugated to a protein this is usually referred to as a glycosylated protein or a glycoprotein. N-linked glycosylation occurs *via* the amide nitrogen of asparagine side chains contained within the motif NXS/T (where X can be any amino acid except proline) and O-linked glycosylation *via* the hydroxyl group of serine and threonine side chains. Glycosylated proteins typically reside on the exterior surface of cells, although they can exist in all cellular locations to varying degrees. Glycosylation of proteins represents one of the major PTMs encountered in eukaryotic systems. Glycosylation plays a role in the biological function, stability, solubility and metabolic rate of the glycosylated protein (from now on referred to as a glycoprotein). Characterization of the oligosaccharide structure aids in interpreting biological processes such as cell-surface recognition, control of protein turnover, intercellular recognition, structural integrity of enzymes and many other functions associated with disease progression.[11–22] A wide heterogeneity exists amongst oligosaccharide chains due to the multiple enzymes involved in assembling,

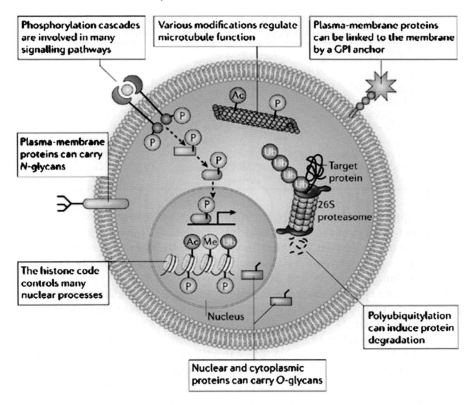

Phosphorylation cascades are involved in many signalling pathways

Various modifications regulate microtubule function

Plasma-membrane proteins can be linked to the membrane by a GPI anchor

Plasma-membrane proteins can carry *N*-glycans

The histone code controls many nuclear processes

Polyubiquitylation can induce protein degradation

Nuclear and cytoplasmic proteins can carry O-glycans

Nucleus

Target protein

26S proteasome

Figure 7.1 Cellular PTMs. This schematic figure shows the location and role of a selection of some of the most important of more than 200 types of PTM. PTMs are found on all types of proteins and affect the physiochemical properties of proteins, which provides a mechanism for the dynamic regulation of molecular self-assembly and catalytic processes through the reversible molecular recognition of proteins, nucleic acids, metabolites, carbohydrates and phospholipids. Ac, acetyl group; GPI, glycosylphosphatidylinositol; Me, methyl group; P, phosphorylation group; Ub, ubiquitin.[29] (Reproduced with kind permission from *Nature Reviews Molecular Cell Biology*.)

trimming and maturation of the carbohydrate chains. This marked diversity of glycan structures is attributed to complex chain branching which can be derived from any number of available monosaccharides. Table 7.1 lists some of the key sugars found in human glycoproteins and their common abbreviated nomenclature. This complexity makes the characterization of carbohydrate moieties extremely challenging and many of the traditional methods for oligosaccharide analysis are slow and cumbersome.[13,23–27]

Microwave-assisted techniques for the characterization of oligo- and monosaccharides (as well as fatty acids and sphingoids) were first reported by Itonoria *et al.*[28] Methanol hydrolysis under basic conditions was performed

Table 7.1 Principal sugars found in human glycoproteins and their common abbreviated nomenclature.

Sugar	Type	Abbreviation
Galactose	Hexose	Gal
Glucose	Hexose	Glc
Mannose	Hexose	Man
N-acetylneuraminic acid	Sialic acid (9 carbon atoms)	NeuAc
Fucose	Deoxyhexose	Fuc
N-acetylgalactosamine		GalNAc
N-acetylglucosamine	Aminohexase	GlaNac
Xylose	Pentose	Xyl

on glycosphingolipids by employing a 2 min microwave reaction to produce a by-product-free lysoglycosphingolipid intermediate. A subsequent 45 s microwave exposure to 1 M HCl in methanol followed by extraction completed the hydrolysis. Using microwave-assisted hydrolysis, this traditionally time-consuming technique was reduced from hours to minutes.

In 2005 Lee *et al.* described a technique for employing microwave-assisted partial acid hydrolysis for the characterization of monosaccharides obtained from glycopeptides.[29] A domestic-grade GE JES1036 turntable microwave oven (General Electric, Louisville, KY) with an output power of 1100 W (power level 10) and a frequency of 2450 MHz was employed. Incubation of standard N-glycosylated peptides in various concentrations of trifluoroacetic acid (TFA) for between 30 and 120 s of exposure was explored which resulted in partial cleavage of the oligosaccharides. The carbohydrate moieties were analyzed using matrix-assisted laser desorption ionization (MALDI) time-of-flight (TOF) mass spectrometry. A mixture of oligosaccharides as well as the partially deglycosylated peptides were observed. Figure 7.2 shows the MALDI-TOF mass spectra of the reaction mixtures from microwave-assisted acid hydrolysis of a glycopeptide of molecular weight 3355 Da derived from horseradish peroxidase (HRP) corresponding to peptide S_{265}–R_{283}.[29]

Overall, the most labile group within the oligosaccharides was observed as the fucose (Fuc) residue and a majority of the end cleavage products were peptides with one N-acetylglucosamine (GlcNAc) residue linked to asparagine (Asn) (with some peptides containing more than one such group). The data also showed examples where a water molecule (−18 Da loss) or an acetyl group (−42 Da loss) was observed for some hydrolyzed products of the carbohydrate moiety of the glycopeptides. The TFA-mediated microwave method reduced hydrolysis times from 1 h for conventional thermal heating to 60 s. This method proved particularly useful in identifying glycopeptides and obtaining monosaccharide compositions of glycopeptides in a rapid manner. Enough cleavage was observed in many cases to obtain monosaccharide composition information from a wide range of glycopeptides.

Lee *et al.* extended this protocol to the analysis of oligosaccharides from intact glycoproteins.[30] Bovine pancreatic ribonuclease B (RNase B, $M_r = 15$

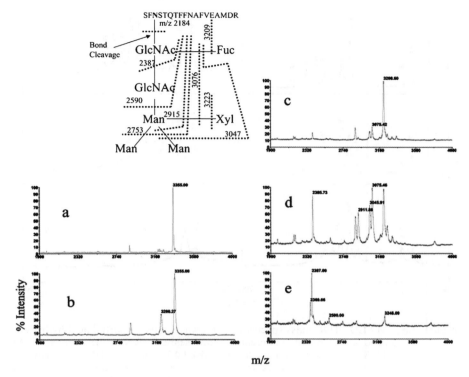

Figure 7.2 MALDI-TOF mass spectra of the reaction mixtures from microwave-assisted acid hydrolysis of a glycopeptide of molecular weight 3355 Da derived from horseradish peroxidase (HRP) corresponding to peptide S_{265}–R_{283} after (a) 14 mM TFA and no microwave exposure, (b) 1.4 M TFA and 60 s of microwave exposure, (c) 3.1 M TFA and 60 s of microwave exposure, (d) 4.5 M TFA and 30 s of microwave exposure and (e) 6.7 M TFA and 60 s of microwave exposure.[29] (Reproduced with kind permission from Wiley.)

kDa), egg white avidin ($M_r = 16$ kDa), human serum a1-acid glycoprotein (a1-AGP, $M_r = 36$ kDa) and fetal calf serum fetuin (BSF, $M_r = 45$ kDa) were employed as model systems which were subjected to limited acid hydrolysis using microwaves to partially cleave oligosaccharides from the whole protein. Monosaccharide structures were obtained for proteins ≤ 20 kDa with MALDI-TOF analysis (with the molecular weight limitation being largely due to the analytical capabilities of the mass spectrometer rather than the microwave-assisted method). Identical conditions were employed as described above for glycopeptides hydrolysis. Figure 7.3 shows an example of this technique employed on the glycoprotein pancreatic ribonuclease B. Here, MALDI-TOF mass spectra of microwave-assisted partial acid hydrolysis products of RNase B are demonstrated. The non-glycosylated RNase B at m/z 13 690 is marked with an asterisk.

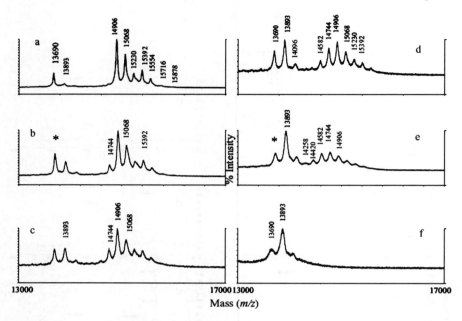

Figure 7.3   MALDI-TOF mass spectra of microwave-assisted partial acid hydrolysis
products of bovine pancreatic ribonuclease B (RNase B) with (a) 1.2 M
TFA and no microwave exposure, (b) 4.4 M TFA and 60 s of microwave
exposure, (c) 4.4 M TFA and 90 s of microwave exposure, (d) 4.4 M TFA
and 120 s of microwave exposure, (e) 6.5 M TFA and 60 s of microwave
exposure and (f) 6.5 M TFA and 120 s of microwave exposure. The
non-glycosylated RNase B at *m/z* 13 690 is marked with an asterisk.[30]
(Reproduced with kind permission from Wiley.)

7.3   Microwave-Assisted Enzyme-Mediated N-linked Deglycosylation

In the method of Lee *et al.* described above, deglycosylation is performed with
the purpose of subsequently characterizing the oligosaccharide moieties. During microwave-assisted TFA hydrolysis the peptide or protein is partially or
completely hydrolyzed. In many cases deglycosylation is performed to reduce
the heterogeneity of a protein or mixture of proteins, in which case an enzyme is
typically employed which removes the oligosaccharide while leaving the peptide
or protein intact. The most common method for the specific removal of
N-linked oligosaccharides is by incubation with the enzyme peptide
N-glycosidase F (PNGase F). PNGase F is commonly employed in the biotechnology industry for the removal of the main G0, G1 and G2 oligosaccharides from Asn 297 of the heavy chain of antibodies (and removal of
N-linked oligosaccharides from other recombinant proteins). Sandoval *et al.*
described a method for the accelerated removal of *N*-linked oligosaccharides
using PNGase F assisted by microwave irradiation. Complete deglycosylation

was achieved in less than 30 min for most proteins without compromising the integrity of protein samples. This method was employed on a variety of glycoproteins, including antibodies, at 10 µg or less.[31] Many parameters including incubation time, temperature and enzyme-to-substrate ratio were explored, and although protein dependent, findings indicated that an incubation temperature of between 37 and 45 °C with between 2 and 5 W of microwave power in a CEM Discover microwave system gave complete deglycosylation for the majority of proteins characterized in under 30 min. Figure 7.4 shows representative deconvoluted mass spectra of reduced Avastin® heavy chain.

Sandoval *et al.* investigated the effects of the addition of enzyme-friendly surfactants such as Rapigest™ or the addition of organic solvent (10–20% ACN) on the rate of microwave-assisted PNGase F catalysis.[31] Under these denaturing conditions in water bath mediated incubations, enzymatic reactions could often be accelerated due to increased accessibility of the active site of the enzyme to its substrate. In the study by Sandoval *et al.*, addition of organic solvent did not show any marked decrease in deglycosylation time; however, addition of Rapigest did greatly decrease reaction time. Addition of 0.1% Rapigest decreased the microwave-assisted reaction time to 10 min, but significant sample losses and precipitation were observed; therefore, for high recovery of low-level material it was decided not to further pursue the use of such surfactants.[31]

One of the main applications that benefited from microwave-assisted enzyme-mediated *N*-linked deglycosylation was the removal of oligosaccharides from conjugated recombinant proteins. Antibodies, and indeed other therapeutic proteins, may be conjugated with bifunctional chelating agents (BCAs), which in turn bind radioactive metal ions, this enabling radioimaging to track the biodistribution of therapeutic molecules in animal models. One example of a BCA is DOTA (1,4,7,10-tetraazacyclododecane-*N*,*N'*,*N''*,*N'''*-tetraacetic acid), a well-characterized reagent which is commonly employed as a visualization tool for biodistribution as it can produce a physiologically stable complex with trivalent radio-metals. When coupling DOTA to biotherapeutic molecules it is important to know how many molecules are conjugated per drug molecule. As the addition of DOTA adds heterogeneity to the protein mass envelope, it is important that additional heterogeneity due to the presence of oligosaccharides is minimized in order for intact molecular weight analysis by mass spectrometry to proceed. *N*-linked deglycosylation with PNGase F on conjugated antibodies is often an order of magnitude slower than deglycosylation of the non-DOTA-modified counterpart. The reason for this is mainly attributed to the lack of accessibility of the enzyme to the oligosaccharide due to steric hindrance caused by the conjugate group. To further test the applicability of microwave-assisted *N*-linked deglycosylation, DOTA-conjugated antibodies were employed to see if these typical 48 h deglycosylation reactions could be reduced significantly as observed for the non-conjugated glycoproteins. Sandoval *et al.* demonstrated that 1 h of microwave-assisted *N*-linked deglycosylation was sufficient for efficient removal of the oligosaccharides without any detrimental effect to the DOTA molecule. Removal of the

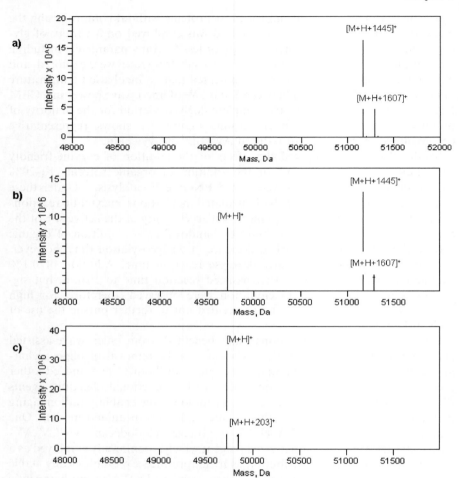

Figure 7.4   Representative deconvoluted mass spectra of reduced avastin heavy chain
(a) prior to deglycosylation where the parent ion + 1445 Da representing
the G0 (Asn 297-GlcNac-GlcNac-mannose-[mannose-GlcNac]$_2$) (in which
GlcNac is *N*-acetylglucosamine) and + 1607 Da representing the mannose
capped G1 variant, (b) after 5 min deglycosylation in the microwave and
(c) after 10 min microwave-assisted incubation at 37 °C.[31]

oligosaccharides allowed the number of DOTA molecules to be visualized (each
with a molecular weight of 386 Da). Figure 7.5a shows a chromatogram and
raw mass spectrum of reduced glycosylated anti-CD4 heavy chain conjugated
to DOTA. The spectrum could not be deconvoluted due to the complex heterogeneity of the sample from the presence of both glycosylation and DOTA
conjugation. Figure 7.5b shows the mass spectrum of the DOTA-conjugated
anti-CD4 antibody after 1 h incubation in the microwave with PNGase F. To
gain the same level of deglycosylation in the water bath also at 37 °C, it was

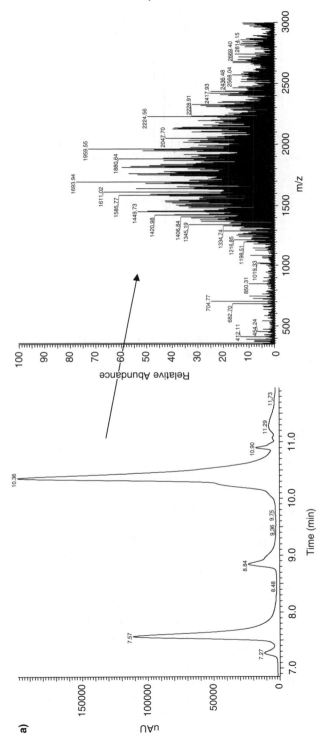

Figure 7.5 (a) Chromatogram and raw mass spectrum of reduced glycosylated anti-CD4 heavy chain conjugated to DOTA. The spectrum could not be deconvoluted due to the complex heterogeneity of the sample from the glycosylation and DOTA conjugation. Mass spectra of the DOTA-conjugated anti-CD4 antibody (b) after 1 h incubation in the microwave with PNGase F and (c) after 24 h incubation in the water bath. Conjugation adds a mass of 386 Da per DOTA molecule to the biomolecule.[32] (Reproduced with permission from Elsevier.)

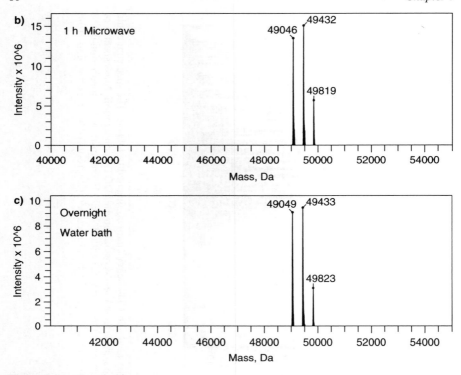

Figure 7.5 *Continued.*

necessary to perform a 24 h incubation; therefore, again, this demonstrated that the microwave-assisted deglycosylation dramatically decreases incubation times for the deglycosylation of N-linked glycoproteins.[31] For a comprehensive protocol for microwave-assisted enzyme-mediated N-linked deglycosylation, refer to protocol VIII.

Tzeng *et al.* further developed the microwave-assisted PNGase F-mediated deglycosylation protocol described by Sandoval *et al.* for the facile MALDI mass spectral analysis of neutral glycans.[32] After performing an in-solution microwave-assisted tryptic digestion followed by microwave-assisted deglycosylation in the presence of PNGase F, they employed carboxylated/oxidized diamond nanoparticles to perform a selective solid-phase extraction to remove the proteins and peptides from the released glycans. To minimize MALDI mass spectral signal suppression effects due to cation adducts such as K$^+$ and Na$^+$, a NaOH solution was mixed with the acidic matrix. In addition to suppressing formation of potassiated and sodiated oligosaccharide ions, this method also suppressed the spectral signal of peptides that had not been fully retained by the diamond nanoparticles. By combining all of the above methods, the analysis of neutral glycans from proteins can now be completed in less than 2 h in contrast to the 2 days typically required using conventional methods.[32]

7.4 Microwave-Assisted O-linked Deglycosylation

In addition to the microwave-assisted enzyme-mediated *N*-linked deglycosylation work discussed above, Sandoval *et al.* also went on to explore the utility of *O*-linked enzyme-mediated microwave-assisted deglycosylation. Employing bovine fetuin as a model protein they demonstrated that traditional water bath mediated incubations provided identical results to microwave-assisted incubation, and that for a variety of *O*-linkage oligosaccharide-removing enzymes, the deglycosylation reactions all typically occurred with ease in a short time frame (<2 h). Therefore, to date, there is no evidence that microwave-assisted *O*-linked enzyme-mediated deglycosylation benefits from microwave-assisted incubations.[33]

A novel method for distinguishing between *O*- and *N*-linked glycosylation sites on proteins was reported by Li *et al.* This group had previously shown that microwave-assisted acetic acid or formic acid cleavage can cleave proteins specifically at aspartic acid residues.[34] In a follow-up to this work they showed how this microwave-assisted method related to the cleavage of glycoproteins and concluded that microwave-accelerated acetic acid treatment provided residue-specific cleavage of the polyamide backbone in *N*-linked glycoproteins, without hydrolyzing the *N*-linkage or the carbohydrate side chain. However, the same treatment provided residue-specific cleavage of the polyamide backbone in *O*-linked glycoproteins; however, the *O*-linkage was hydrolyzed to release the carbohydrate chain.[35]

7.5 Microwave-Assisted Methods for Phosphorylation Mapping

Phosphorylation (typically of serine, threonine or tyrosine residues) is a common PTM in eukaryotic and prokaryotic organisms, which can direct the activity and function of a protein. The progression of many oncological pathways are dictated by kinase activity (the enzymes responsible for phosphorylation events); hence phosphorylation mapping is currently one of the most active areas of proteomic research.[36] There are several methods commonly employed for flagging phosphorylation sites or isolating phosphopeptides including the use of β-elimination and Michael addition, and also immobilized metal affinity chromatography (IMAC) both of which have been further optimized by microwave assistance.

7.5.1 Microwave-Assisted β-Elimination and Michael Addition

Many protocols for mapping phosphorylation sites involve either enrichment of phosphopeptides, or indeed modification with or without subsequent modification, for precise site mapping or characterization of recombinant proteins and phosphorylation sites. One such tool is the β-elimination followed by Michael addition of a thiol substrate to the phosphorylated amino acid (in the case of serine and threonine).[32] This reaction can alter the unstable phosphorylated species into a stable derivative which can then be either sequenced by N-terminal sequencing to pinpoint the modified amino acid, analyzed by tandem mass spectrometry, or, if a thiol containing a capture tag is used during Michael addition, can be selectively enriched from a complex mixture. This reaction can also be performed on *O*-linked glycosylation residues as the same principles apply. For both glycosylation site mapping and phosphorylation site mapping this is a very useful tool for determining the exact location of a phosphorylation site, particularly when more than one serine or threonine residue is present in close proximity.

Sandoval *et al.* investigated an alternative method for the precise mapping of phosphorylation sites involving microwave-assisted β-elimination and Michael addition. Here, elective modifications of phosphorylated residues were converted to the more stable *S*-ethylcysteine (phosphoserine) and β-methyl-*S*-ethylcysteine (phosphothreonine) derivatives.

At high pH, the phosphate ester groups of phosphoserine or phosphothreonine residues are compromised, resulting in an α,β-desaturated moiety corresponding to a neutral loss of phosphoric acid (-98 Da). This process of β-elimination creates a Michael substrate that is susceptible to nucleophilic addition. Though a number of different nucleophiles have been reported, after evaluation of more than twenty compounds for completeness of reaction, derivative visibility by Edman sequencing and ease of use, Sandoval *et al.* selected the alkyl thiols ethanethiol and 1-propanethiol for these reactions. After β-elimination, addition of ethanethiol to the phosphopeptide or protein resulted in a net loss of 36 Da and 1-propanethiol in a net change of -22 Da after successful nucleophile incorporation.[32] Figure 7.6 shows a stepwise schematic of the reaction.

Traditionally such nucleophilic derivatization reactions are performed in a water bath for between 1 and 3 h at 60 °C. Phosphoserine residues readily undergo β-elimination within 1 h, but phosphothreonines typically require longer incubation times for a complete reaction, and even then, often after extended incubations, complete reactions are not observed. Sandoval *et al.* demonstrated that equivalent results to those observed after a 3 h incubation of β-eliminated phosphopeptides/proteins with propanethiol may be obtained by microwave incubation for 2 min at 100 °C. Figure 7.7a shows the spectrum of the start material which was composed of a mixture of phosphorylated and non-phosphorylated peptides. Figure 7.7b shows the spectrum of the peptide solution after the β-elimination reaction at 60 °C in a water bath for 1 h. Both the eliminated (-98 Da) and the ethanethiol-modified moieties (-36 Da) of the phosphothreonine and phosphoserine peptides were observed. The non-phosphorylated peptide peaks (unlabeled) do not differ from the starting material. Figure 7.7c shows the β-elimination/Michael addition reaction performed in the microwave for 2 min at 100 °C. The results observed were comparable to those observed for the water bath mediated incubation. Although the Michael addition reaction was not complete under either sets of conditions (which is a common observation), equivalent results were obtained in a much shorter time using microwave assistance as compared to traditional water bath incubations.[32]

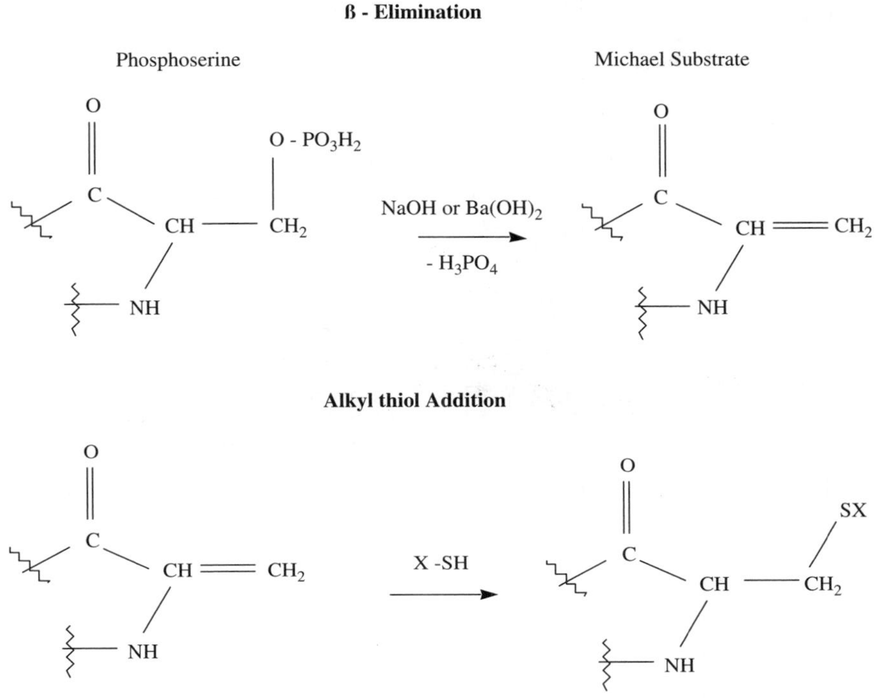

ß - Elimination

Figure 7.6 Stepwise schematic of the β-elimination and Michael addition reaction.

N-terminal sequencing is sometimes employed for phosphopeptide mapping after β-elimination and Michael addition reactions, as it offers a sequential analysis tool to pinpoint phosphorylation sites where multiple serine or threonine exist in the peptide sequence, and which may be ambiguously assigned employing tandem mass spectrometry alone. A convenient sample format for Edman chemistries is to embed a protein in a poly(vinylidine difluoride) (PVDF) membrane. Therefore, Sandoval *et al.* investigated if the rate of reaction for on-membrane β-elimination using microwave assistance could also be improved compared to traditional water bath incubation techniques. Samples were separated by sodium dodecyl sulfate polyacrylamide gel electrophoresis (SDS-PAGE) and blotted onto a PVDF membrane; the bands were then excised from the membrane, wetted with methanol and incubated in the reaction mixture as described above using 1-propanethiol as the nucleophile. Samples were incubated in a microwave for 2 min to 1 h or in a water bath at the same time points. The samples were then washed, dried and sequenced by Edman degradation on a Procise Sequencer (Applied Biosystems, Foster City, CA). After 2 min in the water bath the reaction had not yet commenced. However, after 1 h in the water bath results were found to be comparable to those achieved after just 2 min using microwaves (data not

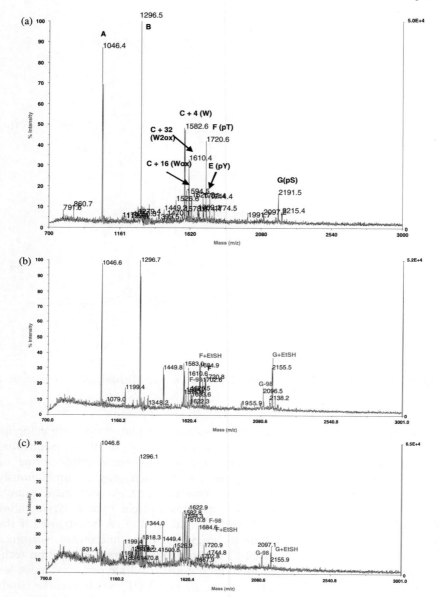

Figure 7.7 (a) Start material consisting of a mixture of phosphorylated and non-phosphorylated peptides. (b) The peptide solution after the β-elimination reaction at 60 °C in the water bath for 1 h. Both the eliminated (−98 Da) and the ethanethiol-modified moieties (−36 Da) of the phosphothreonine and phosphoserine peptides were observed. The non-phosphorylated peptide peaks (unlabeled) do not differ from the starting material. (c) The β-elimination/Michael addition reaction performed in the microwave for 2 min at 100 °C.[32]

shown); peaks representing the modified serine peaks eluted just after leucine on the phenylthiohydantoin chromatogram. This demonstrated that microwave-assisted β-elimination and Michael addition experiments could be performed both in solution and on membrane in a faster and more efficient manner when performed with microwave assistance as compared to traditional water bath incubations.[32]

7.5.2 Enrichment of Phosphopeptides using Microwave-Assisted IMAC Probes

Chapter 4 highlighted the application of magnetite beads for accelerated microwave-assisted enzymatic digestions.[37] These multifunctional magnetite beads accelerated microwave-assisted digestions by their ability to absorb microwave radiation more efficiently than conventional solution-based proteolysis. The beads acted as "trapping probes" whereby the negatively charged functionality of the beads allowed proteins to adsorb to the surface of the beads due to electrostatic attraction, hence inducing an increased surface area of the protein, leading to a concentration effect of the protein near to the microwave-sensitive material. In addition, proteins became denatured and therefore more vulnerable to proteolysis once adsorbed to the bead surface.[29] Chen *et al.* went on to explore the utility of magnetite beads further by coupling zirconia to the beads for phosphopeptide enrichment. IMAC, using iron or gallium, and metal oxide affinity chromatography (MOAC) are common tools for the isolation of phosphopeptides from mixtures of modified and unmodified peptides. The magnetite beads coated with zirconia are not commercially available at this time but their synthesis has been described.[38]

Proteins for phosphorylation characterization were denatured and mixed with the suspension containing magnetic particles coated with zirconia by pipetting up and down for 1 min. After rinsing the particles, trypsin in ammonium bicarbonate was added to the mixture and the sample was heated using a domestic microwave oven (power of 900 W) for 1 min to carry out particle-mediated tryptic digestion. Samples were then acidified, and the resultant peptides were mixed vigorously with the beads by pipetting. By doing so, phosphorylated peptides became adsorbed to the zirconia. Particles were rinsed and phosphopeptides eluted using 0.15% TFA mixed with 2,5-DHB (30 mg mL^{-1}) containing 0.5% phosphoric acid with resultant peptides analyzed by MALDI-TOF. Figure 7.8a shows the direct MALDI mass spectrum of a nonfat milk sample and Figure 7.8b a MALDI mass spectrum of the sample obtained after on-bead microwave-assisted tryptic digestion and zirconia-mediated phosphopeptide enrichment.[35]

7.6 Microwave-Assisted Enzymatic Removal of N-terminal Pyroglutamyl

In the biotechnology industry, the assessment of monoclonality of antibodies is extremely important. This is performed by either mass spectrometry or Edman

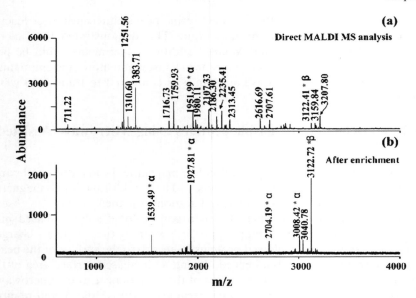

Figure 7.8   (a) Direct MALDI mass spectrum of a nonfat milk sample and (b) a MALDI mass spectrum of the sample obtained after on-bead microwave-assisted tryptic digestion and zirconia-mediated phosphopeptide enrichment.[35] (Reproduced with kind permission from the ACS.)

degradation to guarantee purity, and therefore specificity of the biotherapeutic for its target. A common protocol for determining monoclonality is to perform N-terminal sequencing as far as 30 residues for both the heavy and the light chain. If only a single sequence is observed, monoclonality can be verified; if one or more cycles from Edman degradation contain more than a single residue then the sequence is deemed polyclonal. Murine antibodies expressed in mammalian cell lines are frequently blocked with a pyroglutamyl group where the N-terminal glutamine residue is cyclized by the enzyme glutamine cyclase. This cyclized N-terminal group makes the N-terminus inaccessible to Edman sequencing and in such cases it is necessary to remove the pyroglutamyl group using pyroglutamyl aminopeptidase (PGAP) prior to Edman degradation.

PGAP is a thermostable aminopeptidase used to remove the N-terminal pyroglutamic acid of proteins and peptides and this enzyme is active even at elevated temperatures (as high as 90 °C). A PGAP digestion protocol using a PVDF membrane was recently reported that was performed in a thermocycler at 90 °C for 1 h.[32] Taking advantage of the high thermal stability of this enzyme, the potential of microwave irradiation was explored to determine if microwave assistance offers improved deblocking efficiency over thermocycler-mediated incubations.

To study the efficiency of microwave-assisted PGAP treatment of blocked N-termini, the blocked heavy chain of anti OX-40 ligand antibody was employed. The heavy chain and light chain of this antibody were separated on a

4–20% Tris–glycine gel under reduced conditions. The band corresponding to the heavy chain was excised and treated with Zwittergent 3-16 solution, and the N-terminal pyroglutamyl group was removed by addition of PGAP. Dithiothreitol was added to the digestion buffer to keep the thiol group in the active site of PGAP under reduced conditions in order to maintain the enzyme's activity. PGAP digestion was carried out at 90 °C for 1–60 min in either a thermocycler or a CEM Discover microwave system. The PGAP-treated heavy chain was then subjected to N-terminal sequencing with a Procise Sequencer.[39] Initial yields of the deblocked protein (based on the first five residues), obtained from sequencing data, were calculated. A higher initial yield at each time point indicates faster kinetics of the enzyme.[32]

Figure 7.9 demonstrates the initial yields for the free N-terminus after PGAP treatment of OX-40L heavy chain at each time point past 10 min incubation. The microwave-assisted digestion produced higher initial yields and demonstrated a faster rate of reaction compared to the water bath- or thermocycler-mediated reaction. One hypothesis was that microwave irradiation helped to denature the substrate and allowed more enzyme accessibility; however, for these experiments the protein was already immobilized on PVDF membrane, and therefore accessibility due to protein conformation could be excluded.

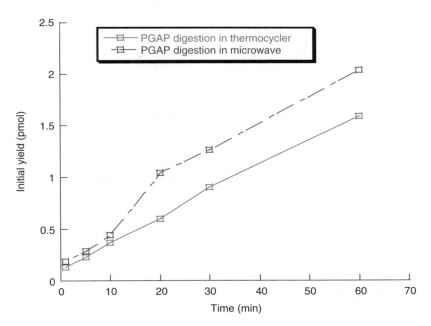

Figure 7.9 Graph showing the initial yields for the free N-terminus after PGAP treatment of OX-40L heavy chain at each time point past 10 min incubation. The microwave-assisted digestion produced higher initial yields and demonstrated a faster rate of reaction compared to the water bath or thermocycler-mediated reaction.[32]

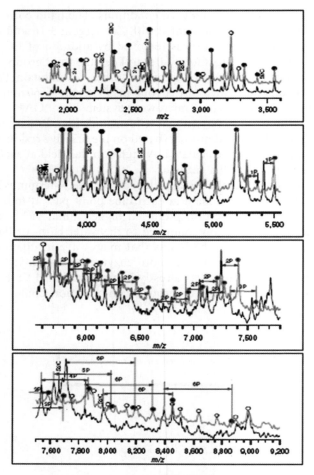

MKLLILTCLVAVALA**RPKHPIKHQGLPQE**
VLNENLLRFFVAPFPEVFGKEKVNELSKD
IGSESTEDQAMEDIKQMEAESISSSEEIVPN
SVEQKHIQKEDVPSERYLGYLEQLLRLKKY
KVPQLEIVPNSAEERLHSMKEGIHAQQKEP
MIGVNQELAYFYPELFRQFYQLDAYPSGA
WYYVPLGTQYTDAPSFSDIPNPIGSENSEK
TTMPLW

Figure 7.10   MALDI spectrum (in red) of a sample containing 85% α-S1-casein and
MALDI spectrum (in black) of a sample containing 85% dephosphorylated α-S1-casein. Peaks labeled with filled circles represent
N-terminal polypeptides and those with open circles the C-terminal
polypeptides. Phosphorylation sites identified are underlined and highlighted in blue.[37] (Reproduced with kind permission from Dr Liang Li.)

A further hypothesis was that the microwave energy may have some effect on the activity of the enzyme itself. PGAP, unlike many other proteolytic enzymes, is heat stable as it is cloned from a thermophilic bacterial species. From the results described by Sandoval *et al.* it is difficult to arrive at a conclusion as to how microwave-assisted PGAP digestions provide more optimal enzyme activity; however, it can be concluded that the removal of N-terminal pyroglutamyl groups by microwave assistance was more efficient than performing the same experiments using conventional heating methods.[32]

7.7   Other Microwave-Assisted Methods for the Characterization of PTMs

During the initial study by Zhong *et al.* using fast microwave-assisted HCl hydrolysis it was observed that some PTMs, *e.g.* phosphorylation sites, were not cleaved under acidic conditions and remained intact for mass spectrometric analysis and mapping.[40] The protein ladder observed in the mass spectral data after limited acid hydrolysis with 1 min of microwave exposure allowed characterization of the amino acid sequence of the protein termini, in addition to the phosphorylation sites of α-casein. A mixture of phosphorylated and dephosphorylated α-casein was subjected to microwave-assisted acid hydrolysis with 6 M HCl, and hydrolyzates were analyzed with MALDI-TOF mass spectrometry. Mass shifts of 80 Da corresponding to the presence of phosphate groups were observed in the mass spectrum at the relevant sites. Additional PTMs could also be observed as being stable under these conditions. Therefore microwave-assisted acid hydrolysis offers yet another beneficial tool, this time for the identification of PTMs from recombinant proteins. Figure 7.10 shows a MALDI-TOF mass spectrum (in red) of a sample containing 85% α-S1-casein and a MALDI spectrum (in black) of a sample containing 85% dephosphorylated α-S1-casein. Peaks labeled with filled circles represent N-terminal polypeptides, and those with open circles the C-terminal polypeptides. Phosphorylation sites identified are underlined and highlighted in blue.

In summary, many of the traditional protocols designed for either the isolation or modification of common protein modifications can be enhanced by microwave-assisted techniques to offer more robust, faster and in some cases alternative methods for the characterization of PTMs.

References

1. R. Uy and F. Wold, *Science*, 1977, **198**, 890.
2. O. Jensen, *Nat. Rev. Mol. Cell Biol.*, 2006, **7**, 391.
3. N. Bandeira, K. Clauser and P. A. Pevzner, *Mol. Cell Proteomics*, 2007, **6**, 1123.
4. N. Bandeira, D. Tsur, A. Frank and P. A. Pevzner, *Proc. Natl Acad. Sci. USA*, 2007, **104**, 6140.

5. G. L. Corthals, R. Aebersold and D. R. Goodlett, *Methods Enzymol.*, 2005, **405**, 66.
6. K. Larsen, M. B. Thygesen, F. Guillaumie, W. G. Willats and K. J. Jensen, *Carbohydr. Res.*, 2006, **341**, 1209.
7. P. Liu, C. L. Feasley and F. E. Regnier, *J. Chromatogr. A.*, 2004, **1047**, 221.
8. R. D. Unwin, J. R. Griffiths, M. K. Leverentz, A. Grallert, I. M. Hagan and A. D. Whetton, *Mol. Cell Proteomics*, 2005, **4**, 1134.
9. S. Mollah, I. E. Wertz, Q. Phung, D. Arnott, V. M. Dixit and J. R. Lill, *Rapid Commun. Mass Spectrom.*, 2007, **21**, 3357.
10. D. N. Perkins, D. J. Pappin, D. M. Creasy and J. S. Cottrell, *Electrophoresis*, 1999, **20**, 3551.
11. P. Hu, P. Berkowitz, V. J. Madden and J. Rubenstein, *J. Biol. Chem.*, 2006, **281**, 12786.
12. X. L. Zhang, *Curr. Med. Chem.*, 2006, **13**, 1141.
13. A. Varki, R. Cummings, J. Esko, H. Freeze, G. Hart and J. Marth (Eds), *Essentials of Glycobiology*, Cold Spring Harbor Laboratory Press, New York, 1999.
14. H. Freeze, *Glycobiology*, 2001, **11**, 129.
15. N. Yonezawa, S. Amari, K. Takahashi, F. L. Imai, S. Kanai, K. Kikuchi and M. Nakano, *Mol. Reprod. Dev.*, 2005, **70**, 222.
16. N. Helenius and M. Aebi, *Annu. Rev. Biochem.*, 2004, **73**, 1019.
17. E. S. Trombetta, *Glycobiology*, 2003, **13**, 77R.
18. E. S. Trombetta and A. J. Parodi, *Annu. Rev. Cell. Dev. Biol.*, 2003, **19**, 649.
19. A. Varki, *Glycobiology*, 1993, **3**, 97.
20. P. M. Rudd, T. Elliott, P. Cresswell, I. Wilson and R. A. Dwek, *Science*, 2001, **291**, 2370.
21. T. Takemoto, S. Natsuka, S. I. Nakakita and S. Hase, *Glycoconjugate J.*, 2005, **22**, 21.
22. J. Zhang, Y. Xie and J. Hedrick, *Anal. Biochem.*, 2004, **334**, 20.
23. C. T. Walsh, *Posttranslational Modification of Proteins*, Roberts and Company Publishers, Englewood CO, 2006.
24. I. Brockhausen (Ed.), *Glycobiology Protocols*, Humana Press, Totowa, NJ, 2006.
25. A. P. Corfield (Ed.), *Glycoprotein Methods and Protocols*, Humana Press, Totowa, NJ, 2000.
26. S. J. Higgins and B. D. Hames (Eds), *Post-Translational Processing*, Oxford University Press, Oxford, 1999.
27. E. F. Hounsell (Ed.), *Glycoanalysis Protocols*, Humana Press, Totowa, NJ, 1998.
28. S. Itonoria, M. Takahashi, T. Kitamura, K. Aoki, J. T. Dulaney and M. Sugita, *J. Lipid Res.*, 2004, **45**, 574.
29. B.-S. Lee, S. Krishnanchettiar, S. S. Lateef and S. Gupta, *Rapid Commun. Mass Spectrom.*, 2005, **19**, 1545.
30. B.-S. Lee, S. Krishnanchettiar, S.-S. Lateef, N. S. Lateef and S. Gupta, *Rapid Commun. Mass Spectrom.*, 2005, **19**, 2629.

31. W. N. Sandoval, F. Arellano, D. Arnott, H. Raab, R. Vandlen and J. R. Lill, *Int. J. Mass Spectrom.*, 2007, **259**, 117.
32. Y.-K. Tzeng, C.-C. Chang, C.-N. Huang, C.-C. Wu, C.-C. Han and H.-C. Chang, *Anal. Chem.*, in press.
33. W. N. Sandoval, V. Pham, E. S. Ingle, P. S. Liu and J. R. Lill, *Comb. Chem. High Throughput Screening*, 2007, **10**, 751.
34. S. Swatkoski, S. C. Russell, N. Edwards and C. Fenselau, *Anal. Chem.*, 2006, **78**, 181.
35. J. Li, K. Shefcheck, J. Callan and C. Fenselau, Proceedings of the 56th Annual American Society for Mass Spectrometry Conference, Denver, CO, 2008.
36. E. S. Witze, W. M. Old, K. A. Resing and N. G. Ahn, *Nat. Methods*, 2007, **10**, 798.
37. W.-Y. Chen and Y.-C. Chen, *Anal. Chem.*, 2007, **79**, 2394.
38. C.-Y. Lo, W.-Y. Chen, C.-T. Chen and Y.-C. Chen, *Proteome Res.*, 2007, **6**, 887.
39. W. J. Henzel, J. Tropea and D. Dupont, *Anal. Biochem.*, 1999, **267**, 148.
40. H. Zhong, Y. Zhang, Z. Wen and L. Li, *Nat. Biotechnol.*, 2004, **22**, 1291.

CHAPTER 8

Recent Microwave-Assisted Applications in the Life Sciences

Abstract

In parallel with the development of microwave assistance in the protein chemistry and proteomics world, other disciplines within the life sciences have also recently explored the capabilities of this mode of catalysis for decreasing reaction times and optimizing reaction rates and efficiencies. This chapter focuses on four technically unrelated topics, all of which have one common feature: the incorporation of microwave assistance. The first part of this chapter briefly describes protein quantitation, firstly by more efficient coupling of commercially available protein/peptide labeling quantitation tools for mass spectrometric analysis (*i.e.* ICAT™ and iTRAQ™ reagents) and secondly by optimizing immunoassay protocols for protein quantitation using metal-enhanced fluorescence detection and chemiluminescence tools.

The second part of this chapter focuses on the introduction of microwave-assisted molecular biology techniques including the amplification of DNA using both microwave-assisted polymerase chain reaction with Taq polymerase, and also incorporation of microwave assistance in the rolling circle DNA amplification technique for the amplification of DNA containing tandem repeat sequences.

The third part of this chapter provides an overview of a microwave-assisted protocol for the characterization of metal-catalyzed reaction sites on proteins whereby controlled microwave exposure was employed to accelerate metal-catalyzed oxidation reactions that site-specifically oxidize copper-bound amino acids in a metalloprotein. In the concluding part, an overview of a method for microwave-assisted antibody–antigen complex dissociation to aid in the analysis of Protein A contaminants in purified biotherapeutic antibody samples is described.

Microwave-Assisted Proteomics
By Jennie Rebecca Lill
© Jennie Rebecca Lill 2009
Published by the Royal Society of Chemistry, www.rsc.org

Although diverse in nature, each of these methods has benefited from microwave assistance and demonstrates the utility of this continually growing field.

8.1 Microwave-Assisted Protein Quantitation Protocols

Multiple tools exist for the quantitation of proteins and complex proteomic samples. These can be divided succinctly into two categories including colorimetric assays (such as fluorescence-based methods, UV methods and chemiluminescence) and physical detection methods such as mass spectrometry. For comprehensive reviews on mass spectrometric quantitation techniques specific for the analysis of proteins, see Lill[1] or Fenselau.[2] Two chemical labeling approaches, *i.e.* ICAT™ and iTRAQ™ labeling, for peptide and protein quantitation have been routinely employed for many in-depth proteomic studies and have recently been optimized by performing the reactions in the presence of microwave radiation.

8.1.1 Microwave-Assisted ICAT™ and iTRAQ™ Reactions for Parallel Proteomic Quantitation

Isotope-coded affinity tags (ICAT™) were first described by Gygi *et al.* and were employed for the accurate quantification and concurrent sequence identification of individual proteins from complex proteomic mixtures.[3] ICAT™ technology comprises a pair of tags consisting of two isotopically labeled sulfhydryl reactive groups, one composed of an eightfold deuterated linker with a biotin affinity tag and the other tag identical expect for a non-isotopically labeled linker (Figure 8.1a). Here, the side chains of cysteine residues from a reduced protein from sample A are reacted with the light version of the ICAT™ reagent, and sample B is treated with the deuterated form of the tag. The two samples are combined, digested with the appropriate enzyme and the biotinylated-tagged (cysteine-containing) peptides are isolated from the complex mixture using an avidin affinity separation. Peptides are released from the affinity resin, separated by reverse-phase chromatography and analyzed by tandem mass spectrometry (Figure 8.1b). A second generation of ICAT™ reagents were recently marketed by Applied Biosystems whereby a cleavable linker was incorporated (to minimize the tag mass addition) and ^{13}C rather than deuterium was employed for the isotopic tag to provide closer reverse-phase chromatography retention times between the light and heavy labeled peptides.

In addition to the cysteine labeling of ICAT™ reagents, another set of isobaric tags, this time that specifically label free amines, are also available for multiplexing quantitation experiments. iTRAQ™ reagents were designed as a set of four isobaric reagents allowing for the identification and quantitation of up to four different samples simultaneously. Whereas ICAT™ is limited to cysteine-containing residues, the iTRAQ™ reagents are more universally applicable and can be added to the free N-terminus and lysine residue of any peptide. The iTRAQ™ reagents were designed as isobaric tags consisting of (1) a charged reporter group that is unique to each of the four reagents, (2) a peptide reactive group and (3) a neutral balance portion. Each of these three components presents an overall mass addition of 145 Da, and, upon fragmentation, each of the four reagents gives rise to one of four unique reporter ions (m/z 114–117) which were selected due to these masses residing in a "quiet" area for typical proteomic tandem mass spectra (*i.e.* no interference from common contaminants or immonium ions). Upon fragmentation, in addition to yielding a strong reporter ion (which is used for quantitation), the reagent also favors generation of strong signature *b*- and *y*-ion series allowing confident identification as well as quantitation.[4] Figure 8.2a demonstrates the structure of these isobaric tags and Figure 8.2b represents a general schematic of a multiplexed reaction comprised of four different samples subjected to iTRAQ™ quantitation.

Rutherford *et al.* explored the utility of performing microwave-assisted labeling of proteins with both the ICAT™ and iTRAQ™ reagents.[5] For the ICAT™ labeling protocol, cleavable ICAT™ reagents were employed and the traditional reaction protocol was compared in parallel to a modified microwave-assisted version which involved performing the labeling reaction of the tag in a Discover microwave unit (CEM) at 50 W maximum power with a maximum temperature of 60 °C with cooling for 10 min. Alternative conditions for the biotin cleavage step using microwave assistance were also explored. The

---

Figure 8.1   (a) Structure of the ICAT reagent. The reagent consists of three elements: an affinity tag (biotin), which is used to isolate ICAT-labeled peptides; a linker that can incorporate stable isotopes; and a reactive group with specificity towards thiol groups (cysteines). The reagent exists in two forms, heavy (contains eight deuteriums) and light (contains no deuteriums).[3] (b) The ICAT strategy for quantifying differential protein expression. Two protein mixtures representing two different cell states have been treated with the isotopically light and heavy ICAT reagents, respectively; an ICAT reagent is covalently attached to each cysteinyl residue in every protein. Proteins from cell state 1 are shown in green and proteins from cell state 2 are shown in blue. The protein mixtures are combined and proteolyzed to peptides, and ICAT-labeled peptides are isolated utilizing the biotin tag. These peptides are separated by micro-capillary high-performance liquid chromatography. The members of a pair of ICAT-labeled peptides are chemically identical and are easily visualized because they essentially coelute, and there is an 8 Da mass difference measured in a scanning mass spectrometer (four m/z units difference for a doubly charged ion). The ratios of the original amounts of proteins from the two cell states are strictly maintained in the peptide fragments. The relative quantification is determined by the ratio of the peptide pairs. Every other scan is devoted to fragmenting and then recording sequence information about an eluting peptide (tandem mass spectrum). The protein is identified by computer searching the recorded sequence information against large protein databases.[3] (Reproduced with permission from *Nature Biotechnology*.)

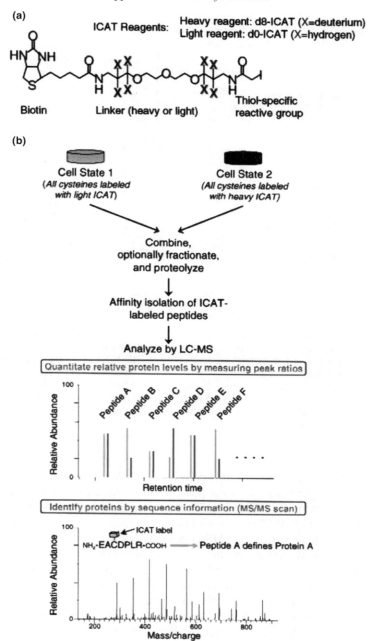

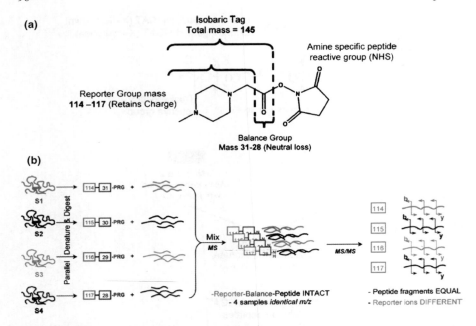

Figure 8.2   (a) The iTRAQ reagent was designed as an isobaric tag consisting of a charged reporter group, a peptide reactive group and a neutral balance portion to maintain an overall mass of 145.[4] (Isobaric, by definition, implies that any two or more species have the same atomic mass but different arrangements.) (b) General scheme of a multiplex reaction of four different samples (S1–S4), designated by four different colors.[4] (Reproduced with permission of Oxford University Press from *Journal of Experimental Botany*.)

total amount of time required to perform the ICAT™ labeling protocol was reduced from more than 10 h to a mere 30 min with equal modification and cleavage of the peptides. The same conditions were also applied for reacting peptides with the iTRAQ™ reagents and here the reaction times were decreased from 2 h to 10 min. Although NHS-esters (the amine chemistry employed for the coupling of iTRAQ™ reagents to free amines) are typically susceptible to degradation upon exposure to high temperatures, surprisingly in this case, there was no detrimental effect from microwave assistance for labeling. This could be due to the very short (30 min) incubation time.

Such vast reductions in incubation times can have a huge impact on the success of detecting low-level proteins as the adsorption of peptides to the wall of the vials and the oxidation of residues such as methionine and tryptophan all increase significantly over a short period of time. By minimizing these effects by introducing shorter incubation times (for digestions as well as for labeling reactions) one can significantly improve on the quality of the proteomic data generated.

8.1.2 Microwave-Assisted Fluorescence and Chemiluminescence Protein Quantitation

Immunoassays are commonly employed for the detection and quantitation of a wide variety of substrates. The typical format of an immunoassay employs antigen–antibody binding for analyte recognition and these most commonly employ fluorescence-based readouts.[6] The two rate-limiting steps of a typical immunoassay are the slow antigen–antibody binding kinetics and the quantum yield of the tagged fluorophore that is used to generate a fluorescent signal readout. Geddes and co-workers published a series of papers on microwave-accelerated protein detection and quantitation methods including microwave-accelerated metal-enhanced fluorescence[6] and chemiluminescence.[7,8] The evolution and utility of these methods are described below. In the first publication of Aslan and Geddes, a combination of metal-enhanced fluorescence (MEF) and low-power microwave assistance was employed to kinetically accelerate assays and to dramatically increase the quantum yield and photostability of weakly fluorescing species.[6]

Conceptually, for microwave-assisted metal-related reactions, the attenuation of microwave radiation arises from the creation of currents from charge carriers being displaced in the electric field.[6] These conductance electrons exhibit extreme mobility and, unlike water molecules, can be completely polarized in extremely short times ($< 10^{-18}$ s). Electrical "arcing" or sparking is not observed so long as the metal particles are sufficiently small or are not present in a concentrated continuous strip. Therefore metallic nanoparticles, fluorophores and microwaves can be employed in combination to yield a kinetically accelerated and optically amplified immunoassay and were shown in preliminary work by Geddes *et al.* to provide a more than ten-fold increase in signal and therefore in assay sensitivity. In addition to this order of magnitude increase in sensitivity, an approximately 90-fold decrease in assay run time was observed. Figure 8.3 shows the benefits of this microwave-accelerated MEF detection system. Here a 30 s incubation, with and without microwave assistance, is shown, demonstrating that significantly greater fluorescein fluorescence emission intensity is observed on silvered surfaces that have been microwave-treated compared to a control sample.

Overall, Geddes *et al.* demonstrated that employment of silver nanostructures can dramatically increase the quantum yield of proximity fluorophores and that by employing low-power microwaves the assay can be rapidly and uniformly heated. The microwaves were proven not to perturb the silver nanostructures or even the proteins being assayed, but instead they simply increased the mass transport of protein to the silvered surface.[6]

A following article by Geddes *et al.* further validated the microwave-accelerated metal-enhanced chemiluminescence method by performing back-to-back comparisons of the method by construction of a model assay sensing platform on both silvered and glass surfaces, and then performing a comparison with identical glass substrate-based assays to confirm the benefit of using silver nanostructures for metal-enhanced chemiluminescence in the presence of

30 sec incubation
No Microwave heating

30 sec incubation
Microwave heating

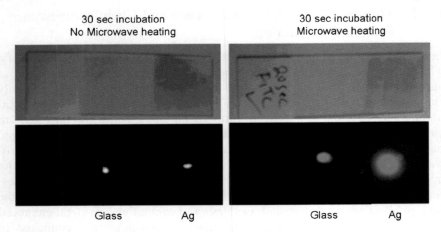

Glass   Ag   Glass   Ag

Figure 8.3 Images showing the benefits of microwave-accelerated metal-enhanced fluorescence. Significantly greater fluorescein emission intensity was observed on the silvered surface that was subjected to microwave heating.[6] (Reproduced with permission from the American Chemical Society.)

microwave radiation.[7] By employing an improved version of their original method (termed microwave-triggered metal-enhanced chemiluminescence, MT-MEC) it was demonstrated that the use of low-power microwaves in combination with enzymes and chemiluminescent species could achieve significantly faster total quantitative protein detection than conventional methods. Figure 8.4 demonstrates the utility of the MT-MEC technique and compares the incubation and reaction times to a traditional Western blot analysis. MT-MEC could detect and quantify surface protein concentrations using low-power microwaves to "trigger" enzyme and chemically catalyzed chemiluminescent reactions due to the "on demand" nature of light emission. This method provided substantial improvements in signal-to-noise ratios from the regular locally triggered reactions.[7]

To further evolve this technology, Geddes *et al.* went on to combine the principles of microwave circuitry and antenna design with their work using MT-MEC to show the potential of "triggering" chemically and enzyme-catalyzed chemiluminescence reactions.[8] Small inexpensive microwave structures which locally accelerate chemical reactions were monitored using finite difference time domain (FDTD). Here, the electric field distribution for various sized and shaped aluminium structures was monitored and bow-tie shaped pieces of aluminium (conceptually modeled on receivers for radiofrequency transmission) were found to be the optimum configuration for dispersed field distribution. Using these aluminium geometric substrates to directionally amplify microwave radiation to accelerate solution-based chemical reactions (*e.g.* chemiluminescence), Geddes *et al.* demonstrated the utility of this microwave-triggered chemiluminescence method for providing another means of dramatically improving signal-to-noise ratios for surface assays especially in the fields of biological assays and chemical sensing technologies.[8]

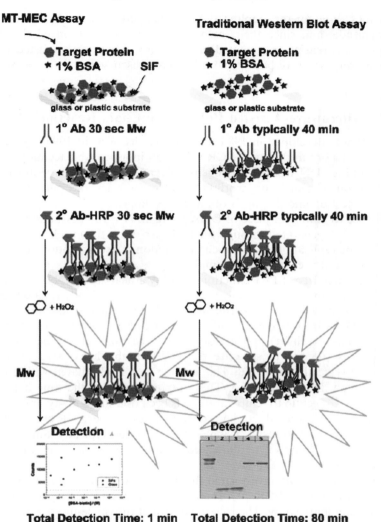

Figure 8.4 Procedure for the MT-MEC immunoassay (Mw, low-power microwave heating).[7] (Reproduced with permission from the American Chemical Society.)

Analogous microwave-accelerated MEF and surface plasmon-coupled directional luminescence methods have also been reported by the same research team specifically for genomic detection applications.[9,10]

8.2 Microwave-Assisted Molecular Biology Tools

Two indispensable tools in molecular biology are the polymerase chain reaction (PCR) and a variation of this technique, rolling circle replication. Both techniques allow for the amplification of nucleotide material by several orders of magnitude, and since their introduction have revolutionized the field of diagnostics, forensics and biomedical discovery. Microwave-assisted versions of these processes have been explored and are discussed in detail below.

8.2.1   Microwave-Assisted Polymerase Chain Reaction

The PCR was developed in 1983 by Kary Mullis, who in 1993 won a Nobel prize for his work in this area. PCR is now an indispensable technique used in DNA cloning, DNA-based phylogeny, diagnosis of hereditary diseases, the identification of genetic fingerprints (used in forensics and paternity testing) and the detection and diagnosis of infectious diseases.[11] PCR employs a DNA polymerase to amplify a piece of DNA using an *in vitro* enzymatic replication. The new DNA generated through the PCR process is then itself employed as a template for replication; hence a chain reaction is set in motion whereby the genetic material is exponentially amplified.

The vast majority of PCR reactions employ a heat-stable DNA polymerase, the most common being Taq polymerase which was originally isolated from the thermophilic bacterium *Thermus aquaticus*. The polymerase assembles the new strand of DNA from free nucleotides which are aligned along a single-stranded DNA template (primer). To perform these reactions at the necessary high temperatures (needed to separate the DNA double-helix strands), thermo-cyclers are typically employed. This heating step is alternated with a cooling step (to allow DNA synthesis). The ramping time, *i.e.* the time it takes the thermocycler to get up to temperature, has been optimized many times over the past decade to allow a reduction in the time necessary per PCR cycle. This slow dissipation of heat transfer using a thermocycler typically limits reaction volumes to 0.2 mL or so, to allow adequate heat penetration in a reasonable amount of time.

In 2003 Fermer *et al.* investigated the utility of microwave irradiation as the source of heat for the PCR.[12] A single-mode microwave cavity was employed for this preliminary study. The lack of control of the single-mode system meant that the critical cooling period could not be applied with microwave control and also there was no means of accurately measuring the temperature within the reaction mixture. Therefore, to compensate, Fermer and colleagues per-formed the microwave treatment in transparent polypropylene tubes which were manually transferred to a temperature controllable heating block for each cycle of primer annealing and the suitable energy content of each irradiation pulse was determined empirically. Both plasmid and chromosomal DNA could be amplified in what were the first of a series of microwave-assisted PCR developments, and even after 25 PCR cycles with microwave assistance, the Taq polymerase remained intact and fully functional.[12]

The same group later went on to further develop the technology and a year later published an efficient method to perform millilitre-scale PCR uti-lizing highly controlled microwave thermocycling.[13] It was demonstrated that high-density microwave *in situ* heating was in many ways superior to traditional heating-block heating as it allows one to avoid large temperature gradients and hot walls of the reaction vessel and therefore provides a more controlled reaction. Using a microwave with an infrared pyrometer to accurately measure and control temperature, a proof-of-principle investigation into controlled microwave-assisted PCR was performed. Microwave-transparent boronic glass tubes were employed for the reaction mixture and a magnetic stirrer was implemented for increased sample and therefore reaction homogeneity. Amplification of a 53 bp fragment from human chromosome 13 was performed with 33 cycles successfully employed after modifications to the method. The optimized method consisted of heating a 2.5 mL sample for 35 s with a target temperature of 88°C, cooling for 50 s and a final heating at 60°C for 85 s. After 33 s a concentration of PCR product between 10 and 30 nM with an amplification efficiency of 92–96% in 1 h and 34 min was obtained.[13] These experiments showed that both Taq polymerase and the nucleotide sequences employed were not destroyed by prolonged microwave irradiation and that PCR reactions can benefit from microwave assistance.

8.2.2 Microwave-Assisted Rolling Circle Amplification

Rolling circle replication describes a process of nucleic acid replication that can rapidly synthesize multiple copies of circular molecules of DNA or RNA, such as plasmids, the genomes of bacteriophages and the circular RNA genome of viroids. It is also thought that some viruses may replicate their DNA *via* a rolling circle mechanism.[14]

In 2006 Yoshimura *et al.* published an article on microwave-assisted rolling circle amplification. The reaction was performed in a volume of 25 μL with 1 μL of primer template mixture. The microwave apparatus is not fully described; however, conditions are cited as a continuous microwave irradiation between 120 and 160 W keeping reaction mixtures at 65°C. Yoshimura *et al.* do not go into much detail other than to suggest that a microwave irradiation PCR rolling circle reaction was more effective than an analogous traditional heating block reaction.[15]

8.3 Microwave-Assisted Characterization of Metal-Catalyzed Reaction Sites on Proteins

To understand the chemistry of a metalloprotein at the molecular level it is first necessary to identify the protein functional groups bound to a transition metal. Although there are several techniques available, each of these has its limitations, and therefore in 2005 Bridgewater and Vachet investigated a mass spectrometric method for determining the localization of copper binding sites on metalloproteins.[16] The method involved the identification of metal-bound amino acids by mass spectrometry *via* metal-catalyzed oxidation (MCO) reactions which were employed to site-specifically oxidize the amino acids coordinated to the metal. When a metalloprotein containing a redox-active metal is exposed to appropriate concentrations of reducing and oxidizing agents, reactive oxygen species are formed which can diffuse from the metal center. These reactive species oxidize neighboring amino acids; therefore mass spectrometry can be employed to analyze proteolytic peptides from the protein to determine metal binding sites.

These reducing/oxidizing reactions using conventional non-microwave-mediated incubations typically take from 30 min to 6 h. Bridgewater and Vachet investigated a microwave-mediated method which demonstrated an increased reaction rate and resulted in incubation times in the range 2–30 min.[16] The main aims of their study were firstly to assess the value of using controlled microwave irradiation to accelerate oxidation of metal-bound residues using this metal-catalyzed reaction and secondly to determine whether the integrity of the structure of the protein is maintained upon microwave irradiation in order for the copper binding site to remain identifiable.

An MDS 2100 microwave oven (CEM) was employed which was operated at atmospheric pressure at microwave settings between 5 and 50% (correlating to microwave powers of between 6 and 411 W). Reactions were performed on the metalloprotein superoxide dismutase (SOD) with reactions initiated by the addition of ascorbate and $S_2O_8^{2-}$. Reactions were terminated by precipitating the protein with acetic acid and trichloroacetic acid. Figure 8.5 shows electrospray ionization mass spectra of the proteolytic digest of Cu/Zn SOD, showing the region around the doubly charged ion of Asp40-Gly49 after a 2 min metal-catalyzed oxidation reaction without and with microwave irradiation at 96 W.

The results from this study demonstrated that controlled microwave irradiation can indeed accelerate the metal-catalyzed oxidation reactions of metal-bound amino acids at least 15-fold. By employing appropriate reagents, oxidation of copper-bound amino acids could be detected in microwave reactions as short as 2 min. However, one caveat to naming this as a high-throughput approach is that for each microwave condition evaluated, a plot of the degree of peptide fragment oxidation as a function of microwave power had to be calculated in order to identify optimum microwave powers for maintaining protein structural integrity. However, this negative point is overshadowed by the improved time resolution for studying dynamic changes to metal binding sites for proteins using this microwave-mediated method as compared to conventional metal-catalyzed oxidation reactions.[16]

8.4   Microwave-Assisted Characterization of Lipase Selectivities

Lipases are a family of water-soluble enzymes that catalyze the hydrolysis of ester bonds in lipid substrates.[17] Lipases are involved in a range of diverse biological processes including routine metabolism and processing of dietary

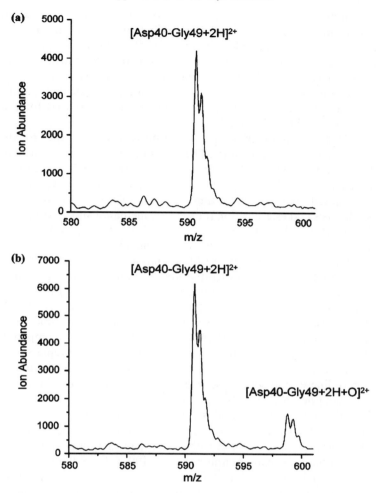

Figure 8.5 Electrospray ionization mass spectra of the proteolytic digest of Cu/Zn SOD, showing the region around the doubly charged ion of Asp40-Gly49 after a 2 min metal-catalyzed oxidation reaction (a) without and (b) with microwave irradiation at 96 W.[16] (Reproduced with permission from the American Chemical Society.)

lipids[18] (*e.g.* triglycerides, fats and other oils) to cell signaling[19] and inflammation.[20] Lipases have a diverse cellular localization *in vivo* and some lipase activities are confined to specific compartments within cells while others work in extracellular spaces.

The history of human exploitation of lipases from fungi and bacteria goes back many millennia, as these enzymes have played roles in the human practice of yogurt and cheese fermentation. In the twentieth and twenty-first centuries, however, lipases are more routinely exploited as cheap and versatile catalysts to degrade lipids in more modern industrial applications. For instance, several recombinant lipases are currently marketed for use in applications such as food processing, laundry detergents and even as biocatalysts in alternative energy strategies to convert vegetable oil into fuel.[21,22] As lipases play such an important role in the catalysis of esterification, transesterification and other applications, it is no wonder that many publications exist demonstrating the necessity to purify, engineer and test the catalysis specificity and speed. Microwave assistance is one of the methods recently explored for the rapid characterization of lipase selectivities.

In 2002 Bradoo *et al.* described the use of microwave assistance for the rapid characterization of lipase selectivities. The rates of reactions for several laboratory-isolated and commercially available lipases were tested and a 7- to 12-fold increase in lipase activity could be demonstrated by the use of microwave irradiation as compared to traditional thermal heating methods.[23] The lipases were evaluated for triolein hydrolysis and esterification (of sucrose, methanol and ascorbic acid) with a selection of fatty acids. All reactions were carried out at 1.35 kW for 30 s. After testing the effect of microwave irradiation on the lipase selectivity it was shown that this remains the same regardless of the catalysis method. The increased lipase activity can equate to higher throughput lipase characterization, and therefore a significantly higher number of characterizations on large enzyme samples and their substrates can potentially be performed in a short amount of time. For the reactions demonstrated by Bradoo *et al.*, a conventional domestic microwave oven was employed and enzymatic reactions were shown to be highly compatible, even in the presence of organic solvents such as hexane. Roy and Gupta summarized the types of reactions catalyzed by lipases in non-aqueous media and also the application of lipases and some other related enzymes which can be used in low-water-containing organic solvents (see Tables 8.1 and 8.2).[24]

Enzymes that can perform catalysis in non-aqueous media are often highly compatible with microwave assistance. For example, enzymes that are compatible with organic solvents tend to be extremely thermally stable, and there are many examples where maintaining them at 100 °C for extended time periods does not cause their inactivation.[25] The hypothesis behind this thermostability in non-aqueous environments (and therefore compatibility with microwave-assisted catalysis) is attributed to the fact that during lyophilization, desiccation removes water molecules which were hydrogen bonded to many surface residues resulting in the side chains of the enzyme creating a rigid structure. Once such lyophilized enzymes are reconstituted in aqueous media, this process is reversed. Therefore if these reactions are carried out in non-aqueous media, this highly conserved tight enzyme cleft is retained; activity is heightened and retained even in the presence of high temperatures, such as those observed using non-temperature controlled microwave assistance.[24,26]

8.5   Dissociation of Protein Complexes

Many biotechnological drug products are composed of monoclonal antibodies which are typically isolated from cell lysates and expression systems using

Table 8.1   Applications of enzymes in non-aqueous media.[24] (Adapted, and with copyright permission, from *Current Sciences*.)

Application	Enzyme	Solvent
Production of biosurfactants	Lipase	Pyridine
Production of biodiesel	Lipase	1,4-Dioxane
Synthesis of aspartame precursor	Thermolysin	Ethyl acetate
Synthesis of (S)-naproxen ester prodrug	Lipase	Isooctane
Synthesis of "delicious" octapeptide	Papain	Acetonitrile
Synthesis of a flavor compound, cis-3-hexen-1-yl acetate	Lipase	Hexane
Production of confectionery fats	Lipase	Solvent-free system
Synthesis and modification of polymers	Lipase, protease, cellulose	Different solvent and co-solvent systems
Oxidation of benzylamine	Amine oxidase	Different organic solvents
Transphosphatidylation of alcohols	Phospholipase D	Chloroform
Synthesis of β-amino acids and β-lactams	Lipase	Diisopropyl ether
Synthesis of non-ionic surfactants	Lipase, protease	Different organic solvents
Acylation of secondary amines	Lipase	Hexane
Acylation of flavonoid (naringin)	Lipase	2-Methy-2-butanol
Preparation of chiral amides	Lipase	Hexane

Protein A affinity purification. This approach yields high purity and high yield; however, the drawback is that some of the Protein A can leach into the resultant drug product. As Protein A is immunogenic and potentially mitogenic, one has to monitor the level of Protein A that is leaching from the Protein A column to ensure that final drug products are Protein A free. Recently, Zhu-Shimoni *et al.* developed and compared two enzyme-linked immunosorbent assay (ELISA) formats for measuring the amount of Protein A leached from the immunoaffinity resin.[27] The samples being tested for trace levels of Protein A contain a vast excess of the biotherapeutic IgG molecule, most of which is present in a Protein A/IgG complex. Although Protein A binding to the IgG molecule is primarily *via* the Fc region, there are also known binding sites in the Fab region. These IgG/Protein A interactions interfere with the binding of the antibody employed for the ELISA assay. In addition, the many different IgG antibody molecules tested for Protein A leaching have different affinities for Protein A and therefore inhibit the detection by ELISA to various degrees. An investigation was therefore launched to overcome these two obstacles.

A method using microwave-assisted dissociation of the IgG/Protein A complex was developed and implemented, initially using a Discover microwave

Table 8.2   Types of reactions catalyzed by lipases in non-aqueous media that
benefit from microwave assistance.[24] (Reproduced with permission
from *Current Sciences*.)

Hydrolysis

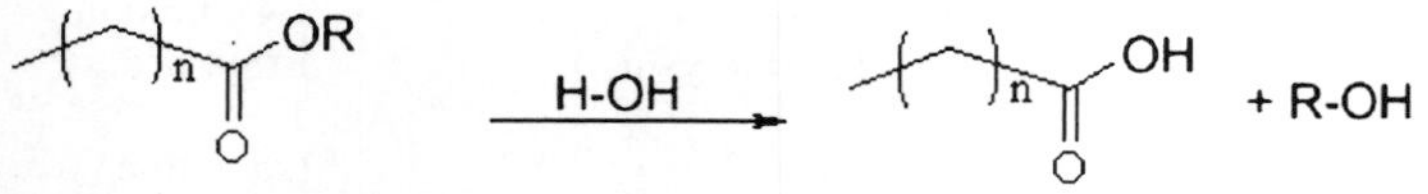

Esterification

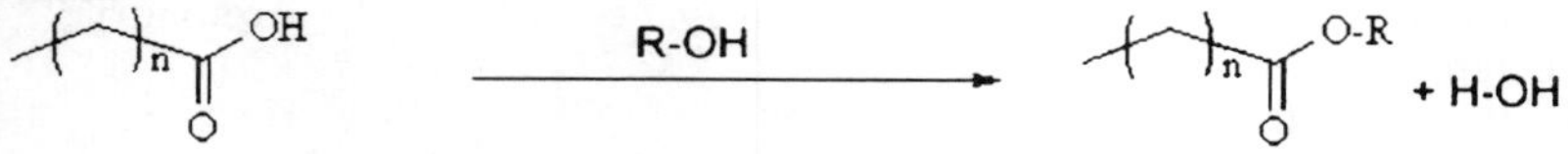

Transesterification

Interesterification

Alcoholysis

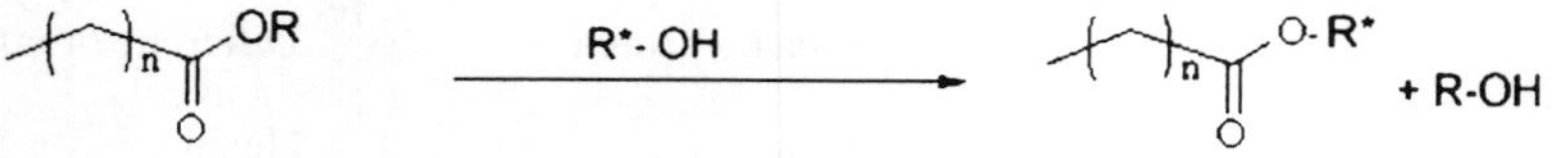

Acidolysis

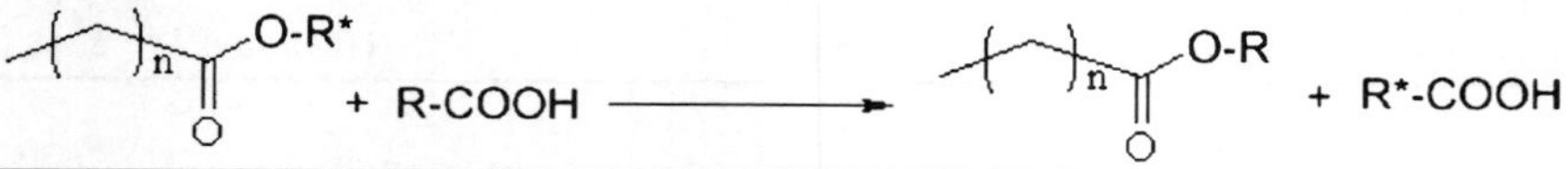

system.[17] A microwave-assisted method was explored because conventional
heating methods induced protein degradation due to long incubation times;
therefore an alternative shorter incubation approach was desirable. In addition,
the dissociation reactions were typically low throughput using conventional
thermal methods. Samples were diluted with a Protein A/IgG dissociation
buffer with a ratio of 1:5 in a 96-well plate prior to microwave heating. Plates
were briefly centrifuged and 5 μL was transferred into 95 μL of dissociation
maintenance assay diluent buffer, after which an ELISA was performed to
measure the amount of leached Protein A. After initial evaluation of the
microwave-assisted dissociation method with the Discover system, Protein
A/IgG dissociation was shown to be effective after just 2 min of microwave
exposure. Figure 8.6 shows the ELISA detection of Protein A in the presence of
antibodies rhMAb1 and rhMAb2 with and without dissociation using microwave-assisted heating.[17]

Dissociation was observed for all three antibodies after just 2 min; however,
throughput using the Discover system was still not high enough and therefore a
MARS model microwave which was adapted to hold three 96-well microtitre

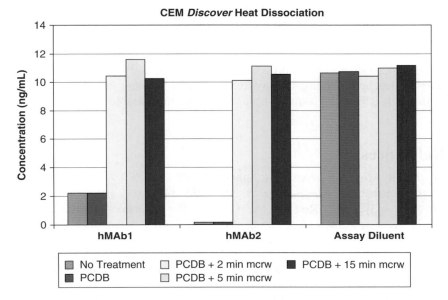

Figure 8.6 Detection of Protein A in the presence of rhMAb1 and rhMAb2, with or without dissociation using microwave-assisted heating. Protein A was used at $12.5\,ng\,mL^{-1}$. (Reproduced with permission from Zhu-Shimoni *et al.*,[27] Genentech Inc., San Francisco, CA.)

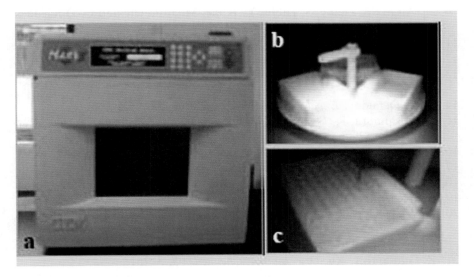

Figure 8.7 CEM MARS microwave instrument: (a) exterior image; (b, c) interior views showing the carousel designed to hold 96-well plates with the fiber-optic probe shown for accurate temperature gauging. (Reproduced with permission from Zhu-Shimoni *et al.*,[27] Genentech Inc., San Francisco, CA.)

plates was employed (see Figure 8.7). Although the MARS system is not as flexible in terms of temperature control and pressure readout as the Discover unit, it did produce the necessary reaction temperatures (90°C) for dissociation to occur.

This microwave-assisted protein dissociation method has been incorporated into the leached Protein A assay at Genentech Inc. and has shown universal utility as it is both high throughput and fully automatable.[17]

References

1. J. Lill, *Mass Spectrom. Rev.*, 2003, **22**, 182.
2. C. Fenselau, *J. Chromatogr. B*, 2007, **855**, 14.
3. S. P. Gygi, B. Rist, S. A. Gerber, F. Turecek, M. H. Gelb and R. Aebersold, *Nat. Biotechnol.*, 1999, **17**, 994.
4. L. R. Zieske, *J. Exp. Botany*, 2006, **57**, 1501.
5. J. L. Rutherford, J. Bonapace, M. Nguyen, T. Pekar, D. Innamorati and J. Pirro, *Proceedings of the CHI Beyond Genome Conference*, San Francisco, CA, 2004.
6. K. Aslan and C. D. Geddes, *Anal. Chem.*, 2005, **77**, 8057.
7. M. J. Previte, K. Aslan, S. N. Malyn and C. D. Geddes, *Anal. Chem.*, 2006, **78**, 8020.
8. M. J. Previte, K. Aslan and C. D. Geddes, *Anal Chem.*, 2007, **79**, 7042.
9. K. Alsan, M. J. Previte, Y. Zhang and C. D. Geddes, *J. Immunol. Methods*, 2008, **331**, 103.
10. K. Alsan, Y. Zhang, S. Hibbs, L. Baille, M. J. Previte and C. D. Geddes, *Analyst*, 2007, **132**, 1130.
11. H. Vanguilder, K. Vrana and W. Freeman, *Biotechniques*, 2008, **44**, 619.
12. C. Fermer, P. Nilsson and M. Larhed, *Eur. J. Pharm. Sci.*, 2003, **18**, 129.
13. K. Orrling, P. Nilsson, M. Gullber and M. Larhed, *Chem. Commun.*, 2004, 790.
14. V. V. Demidov, *Expert Rev. Mol. Diagn.*, 2002, **2**, 542.
15. T. Yoshimura, K. Nishida, K. Uchibayashi and S. Ohuchi, *Nucleic Acids Symp. Ser.*, 2006, **50**, 305.
16. J. D. Bridgewater and R. W. Vachet, *Anal. Chem.*, 2005, **77**, 4649.
17. A. Svendsen, *Biochim. Biophys. Acta*, 2000, **1543**, 223.
18. F. K. Winkler, A. D'Arcy and W. Hunziker, *Nature*, 1990, **343**, 771.
19. S. Spiegel, D. Foster and R. Kolesnick, *Curr. Opin. Cell Biol.*, 1996, **8**, 159.
20. L. W. Tjoelker, C. Eberhardt, J. Unger, H. L. Trong, G. A. Zimmerman, T. M. McIntyre, D. M. Stafforini, S. M. Prescott and P. W. Gray, *J. Biol. Chem.*, 1995, **270**, 25481.
21. Z. Guo and X. Xu, *Org. Biomol. Chem.*, 2005, **3**, 2615.
22. R. Gupta, N. Gupta and P. Rathi, *Appl. Microbiol. Biotechnol.*, 2004, **64**, 763.
23. S. Bradoo, P. Rathi, R. K. Saxena and R. Gupta, *J Biochem. Biophys. Methods*, 2002, **51**, 115.

24. I. Roy and M. N. Gupta, *Curr. Sci.*, 2003, **85**, 1685.
25. A. Zaks and A. M. Klibanov, *Science*, 1984, **224**, 1249.
26. T. Arakawa, S. J. Prestelski, W. C. Kenney and J. F. Carpenter, *Adv. Drug Deliv. Rev.*, 2001, **46**, 307.
27. J. Zhu-Shimoni, F. Gunawan, A. Thomas, J. Stults and M. Vanderlaan, *Poster presented at the Well Characterized Biotechnology Pharmaceuticals meeting*, Washington, DC, 2008.

CHAPTER 9

Epilogue: To Microwave or Not To Microwave?

Although a field of rapid growth, microwave-assisted biological and bio-chemical methods are still in their infancy. There are still multiple questions unanswered as to the exact mechanisms of action of these microwave-assisted methods as compared to traditional heating protocols and the full utility and potential of this emerging field is yet to be realized. At this time, the kinetics and specificity of microwave-assisted incubations and reactions in a proteomic context have only been examined in a very small number of areas and on a limited number of proteins.

Throughout this book multiple methods have been described which benefit from microwave-assisted heating and catalysis. Many researchers will have already formed an opinion on whether or not microwave-assisted methodologies would benefit their individual laboratories. This epilogue is aimed at helping summarize the criteria for spending the time and investment in both personnel and instrumentation for evaluating microwave-assisted methods with the aim of increasing throughput and potentially yield in a proteomic or protein chemistry laboratory.

Traditionally, the need for microwave-assisted reactions was to increase the yield and speed of chemical syntheses. For protein chemists and biochemists, an increase in throughput and efficiency of reaction is also important; however, several different technical considerations have to be posed before embarking on microwave-assisted methods. The main limitations of microwave-assisted protein chemistry reactions are adverse affects on heat-labile enzymes or sub-strate proteins which may either render the enzyme inactive or may induce precipitation of the protein. Microwave-induced enzyme inactivity can be determined if a proteolytic reaction has not occurred, and in addition there is an absence of autolytic site-specific proteolysis peaks from the enzyme of interest. For example, Vesper *et al.* examined the potential of microwave-mediated Glu-C proteolytic digestions (as discussed in Chapter 4); however,

Microwave-Assisted Proteomics
By Jennie Rebecca Lill
© Jennie Rebecca Lill 2009
Published by the Royal Society of Chemistry, www.rsc.org

these were shown to yield fewer proteolytic products than the conventional convection heating method. Autolytic peaks from the Glu-C enzyme were not detected; therefore it was concluded that inactivation of the enzyme was due to microwave-induced denaturation resulting from the instability of this enzyme at elevated temperatures and not due to autolysis.[1]

Precipitation of the protein can occur especially for reactions carried out at higher temperatures under harsher conditions, an example of this being microwave-assisted acid hydrolysis for peptide mass fingerprinting[2] or N-terminal sequencing[3] as discussed in Chapter 6. Precipitation can decrease the rate of reaction due to the tight conformation of the protein structure, rendering it less accessible to the reaction media. In addition precipitation can lead to analytical challenges, especially if employing an online chromatographic separation step. Precipitation can lead to clogged chromatography systems and nanospray tips. To alleviate this problem it is recommended that all samples suspected of precipitation are sieved through a narrow-pore filter or are centrifuged to pellet the precipitate, with the final eluant removed for loading onto the analytical instrumentation using a gel loading pipette tip prior to analysis.

A final consideration is which microwave instrumentation to employ for carrying out microwave-assisted proteomic reactions. Chapter 2 covered most of the advantages and disadvantages of both domestic and laboratory-specific systems. The only disadvantage to investment in a laboratory-specific microwave is the price; however, one has to weigh the benefits of time saving and throughput compared to price.

After reviewing the chapters in this book, the reader/researcher should be armed with enough information to tackle investigations into microwave-assisted proteomics and hopefully the summary of methodologies described from the current literature will ignite some ambition within researchers interested in trying these protocols in their own laboratories.

References

1. H. W. Vesper, L. Mi, A. Enada and G. L. Myers, *Rapid Commun. Mass Spectrom.*, 2005, **19**, 2865.
2. H. Zhong, S. L. Marcus and L. Li, *J. Am. Soc. Mass Spectrom.*, 2005, **16**, 471.
3. H. Zhong, Y. Zhang, Z. Wen and L. Li, *Nat. Biotechnol.*, 2004, **22**, 1291.

CHAPTER 10

Microwave-Assisted Proteomic Protocols

Abstract

This chapter is a synopsis of protocols derived both from the literature as well as from the experience within our laboratory. They are designed for use in both open cavity domestic microwave ovens, or where stated, in industrial based microwave systems such as the CEM Discover microwave unit.

Safety Considerations

1. Wear appropriate protective laboratory clothing and safety glasses at all times.
2. To minimize the risk of microwave exposure, the microwave oven should be turned off and unplugged from the power after each use. In the case of microwave oven malfunction that may result in continued radiation generation even after it is switched off, unplugging the unit from the power should prevent any potential accident from such failure.
3. Take care in handling acids and possible acid vapors during and after microwave hydrolysis. It is recommended to allow the sample vial to cool inside the microwave oven prior to removal and then to open the cooled sample vial in a well-vented bench area, such as inside a fume-hood.
4. Remember, super heating of solvents may occur and may be much hotter than anticipated!
5. For a safety review on handling microwave irradiation in the lab please refer to A. Lew, P.O. Krutzik, M.E. Hart and A.R. Chamberlin,

---

Microwave-Assisted Proteomics
By Jennie Rebecca Lill
© Jennie Rebecca Lill 2009
Published by the Royal Society of Chemistry, www.rsc.org

Increasing Rates of Reaction Microwave-Assisted Organic Synthesis for Combinatorial Chemistry, *J. Comb. Chem.*, 2002, **4**, 95–105.

Protocol I. Microwave Fixing, Staining and Destaining of SDS-PAGE Gels

Reagents

Fixing Reagent
1:2 acetic acid: methanol
Coomassie
250 mL of Ethanol
2.5 g of Coomassie Blue R-250
QS to 1L with SQ H_2O
De-stain
62.5 mL of Acetic Acid Glacial
200 mL of Reagent Alcohol or Ethanol
QS to 1L with SQ H_2O.

Protocol

1. In a domestic microwave place gel/membrane for fixing in a microwave-proof container (Petri dishes are ideal).
2. Add enough fixing reagent to cover the entire gel and ensure the gel is suspended in the fixing reagent and not stuck to the bottom of the dish.
3. Place in a domestic microwave oven and turn on power at full power for 30 seconds.
4. Remove dish and place on shaker for at least 5 min
5. Carefully pour off fixing reagent.
6. Add enough of the Coomassie blue stain to cover the gel/membrane and again ensure the gel is not stuck to the bottom of the dish.
7. Microwave on full power for 1 min.
8. Remove and place on shaker for 1 h at room temperature.
9. Look at the bands on the gel, if needs further staining, repeat procedures 7 and 8.
10. To destain, pour off Coomassie stain and rinse in milli-Q water until water runs clear.
11. Add enough destain solution to cover gel/membrane and again ensure gel is not stuck to bottom of dish.
12. Microwave on full power for 1 min and place on shaker for 30 min.
13. Note: a lint free tissue can be added to increase the de-stain efficiency. Wearing gloves (to minimize keratin contamination) tie the tissue into a knot and place in destain solution next to the gel.

14. After 30 min check the gel, if requires further destaining repeat steps 12 and 13.

Protocol II. Microwave-Assisted Antigen Retrieval

Protocol

1. Wash slides with deionized water, place them in a microwavable container and immerse slides in the antigen retrieval solution of choice or 10 mM sodium citrate pH 6.0 + 1 mM EDTA.
2. Using a domestic microwave-oven, turn on high power (> 600 W) for 5 min and ensure that throughout microwave exposure that the slide remains immersed in the antigen retrieval solution.
3. Repeat process.
4. Cool to room temperature and continue with immunohistochemical staining.

Protocol III. Microwave-Assisted In-Solution Tryptic Digestion

Reagents

Trypsin (recommended Promega sequence-grade modified porcine trypsin, Cat. #V511A)
Ammonium Bicarbonate 25 mM
High quality HPLC grade or MilliQ water

Protocol

1. Dilute trypsin at recommended concentration (1:200 enzyme:substrate) in 25 mM ammonium bicarbonate to sample.
2. Recommended to use an industrial microwave such as the Discover unit by CEM. Add vials/eppendorfs into holder, place into the cavity and program the microwave for 30 min at 5 W at 55 °C with a 2 min ramp time. Domestic microwave oven can be employed, however conditions may vary. If employing a domestic microwave oven its suggested to place a beaker of cold water within the cavity and replace every 5 min.
3. Test an aliquot of sample using either MALDI-TOF MS or SDS-PAGE to check completion of digestion.
4. As trypsin will degrade rapidly at higher temperatures, if digestion is in complete add further aliquot of trypsin and repeat, or for future experiment add small percentage (< 20%) of organic such as acetonitrile or mass spectrometry friendly surfactant (e.g. Rapigest™)

Protocol IV. Microwave-Assisted In-Gel Tryptic Digestion

(For useful overview of in gel enzymatic digestions in general) Bernhard Granvogl, Matthias Plöscher and Lutz Andreas Eichacker. Sample preparation by in-gel digestion for mass spectrometry-based proteomics, *Analytical and Bioanalytical Chemistry*, 2007, **389 (4)**, 991–1002.

Reagents

Trypsin (recommended Promega sequence-grade modified porcine trypsin, Cat. #V511A)
Ammonium Bicarbonate 25 mM
High quality HPLC grade or MilliQ water
HPLC grade Acetonitrile
TFA

Protocol

1. Excise gel band and cut into 4-8 pieces.
2. Wash gel pieces in 25 µL of 25 mM ammonium bicarbonate followed by 25 µL of 25 mM ammonium bicarbonate in 50:50 acetonitrile:water.
3. Dehydrate gel band in 100% acetonitrile, then rehydrate with 0.2 µg of trypsin in 25 mM ammonium bicarbonate (25 µL or enough volume to rehydrate the gel pieces). For best results allow gel to rehydrate on ice for 30 min prior to incubation.
4. Incubate in the microwave at 5 W, 50 °C for 15 min
5. Extract from the gel with 30 µL of 50:50 acetonitrile: water followed by 30 µL of 100% acetonitrile. Pool these extracts. If gel bands are not completely dehydrated (white and crisp) after this step repeat.
6. Reduce volume of organic (preferably by speed-vac) and reconstitute in appropriate buffer for assay.

Protocol V: Microwave-Assisted Protein Hydrolysis for Amino Acid Analysis

Please note: CEM sells a hydrolysis kit with full instructions on how to perform protein hydrolysis for protein quantitation by amino acid analysis. The protocol below can be adapted to a domestic microwave, or, preferably an industrial microwave such as the Discover system from CEM.

Reagents

1. 6 N HCl (analytical grade)
2. Phenol
3. Microwavable vessels (preferably glass)
4. A nitrogen source
5. A vacuum source

General Considerations

- Protein samples should be as pure as possible and any salts or buffers which contain Tris or amines should be avoided as these can react with reagents employed for amino acid labeling subsequent to the hydrolysis step.
- In addition, buffers containing amino acids should be avoided as these can skew quantitation.
- Buffers with high sugar content should also be avoided as these tend to burn and discolor during hydrolysis, leading to incomplete hydrolysis

Protocol

1. Start with a protein concentration and amount suitable for the subsequent amino acid analysis assay.
2. Take sample of interest down to dryness in a speedvac.
3. To sample add 20 μL of 6 N HCl + 0.1% phenol (to minimize oxidation of hydroxyl side chain amino acids such as Ser and Thr).
4. Alternatively purge the tubes with Nitrogen then evacuate the tubes with a vacuum. (If using the CEM vapor phase hydrolysis accessory set instructions will be provided on how to perform this, if the kit is not accessible, please refer to A. Higbee, S. Wong, W.J. Henzel, *Automated sample preparation using vapor-phase hydrolysis for amino acid analysis. Anal Biochem.*, 2003, **318(1)**, 155–158, for instructions on vapor phase evacuation protocols)
5. Perform hydrolysis at 175 °C for 10 min at 300 W, or if using a domestic microwave, perform hydrolysis for 15 min at highest temperature and wattage setting.
6. Dry samples down to completion and complete analysis of hydrolysate as per instructions of the amino acid analysis system employed.

Protocol VI: TFA Cleavage for Bottom-Up Characterization

Reagents

Analytical grade Trifluoroacetic acid

Protocol

1. Reduce and alkylate sample if desired
2. Dilute sample in 25% TFA (in Milli-Q or HPLC grade water)
3. Microwave sample either in domestic microwave oven for 5 min for simple protein mixtures, or 10 min for membrane proteins or complex mixtures
4. If using an industrial microwave employ the following conditions

   - 50 W power
   - 100 °C
   - 1 min ramp time
   - 5 min hold time or 10 min for membrane or complex samples
   - Employ Power-max option if available (cooling option)

5. Remember, if analyzing sample by mass spectrometry remember to include oxidation of methionine and deamidation of Asn and Gln as variable modifications in the search parameters.

Protocol VII: Aspartic Acid Specific Cleavage for Formic Acid Catalysis

Reagents

TCEP
Formic Acid (HPLC grade) or acetic acid (HPLC grade)
Glass microwave-inserts or 96 well plate

Protocol 1 – In Solution

1. To > 500 fmol of protein dilute in 2% Formic acid or 12% acetic acid in the presence of 10 mM TCEP
2. Incubate in the microwave at 50 W, 100 °C for 15 min
3. Cleavage can occur at both the N- and C-terminal side of Aspartic acid.

Protocol 2 – In gel

1. Excise gel band and cut into 4–8 pieces.
2. Wash gel pieces in 25 μL of 25 mM ammonium bicarbonate followed by 25 μL of 25 mM ammonium bicarbonate in 50:50 acetonitrile:water.
3. Dehydrate gel band in 100% acetonitrile, then rehydrate in enough 2% Formic acid or 12% acetic acid in the presence of 10 mM TCEP to cover the gel pieces

4. Incubate in the microwave at 50 W, 100 °C for 15 min
5. Extract from the gel with 30 µL of 50:50 acetonitrile: water followed by 30 µL of 100% acetonitrile. Pool these extracts. If gel bands are not completely dehydrated (white and crisp) after this step repeat.
6. Reduce volume of organic (preferably by speed-vac) and reconstitute in appropriate buffer for assay.

Protocol VIII: *N*-linked Protein Deglycosylation

Reagents

50 mM Tris HCl, pH 7.5
DTT
PNGase F (Preferably Sigma P7367)
1 M Trifluoroacetic acid (TFA)
500 µL insert tubes or eppendorf tubes

Protocol

1. Dilute and reduce sample to a final concentration of 0.5-1.0 mg/mL in 50 mM Tris pH 7.5 containing 50 mM DTT
2. Incubate in microwave at 40 °C at 5 W for 5 min or at room temp for 30 min
3. Add 1–2 µL (0.5–1 Unit) of PNGase F to 20 µL of reduced sample in eppendorf or insert tube
4. Place in chamber of microwave using the following conditions

   - Power 2 W
   - Ramp 0 min
   - Hold time 30 min
   - Temp 35 °C (Actual temp will be approx. 40 °C)
   - Zero pressure

5. Acidify to stop digestion with 2 µL of 3% TFA
6. Analyze sample.
7. If sample not fully de-glycosylated add 10% acetonitrile and increase incubation time by 30 min (to 1 h total).
8. If deglycosylating a complex sample or a heavily modified protein it is recommended to double the amount of PNGase F and increase incubation time to 2 h.

Subject Index

RETURN TO: CHEMISTRY LIBRARY

100 Hildebrand Hall • 510-642-3753

LOAN PERIOD 1	2 1 MONTH	3
4	5	6

ALL BOOKS MAY BE RECALLED AFTER 7 DAYS.

Renewals may be requested by phone ~~or using GLADIS, type~~ **inv** ~~followed by your patron ID number.~~

DUE AS STAMPED BELOW.

FEB 27		

FORM NO. DD 10
3M 7-08

UNIVERSITY OF CALIFORNIA, BERKELEY
Berkeley, California 94720–6000